The AI Developer's Field Guide

Classes, Monsters, and Anti-Patterns in AI Coding

Tim O'Brien

2026

Copyright

The AI Developer's Field Guide

by Tim O'Brien

Contributor: Chris Undernehr

Printed in the United States of America.

Second Edition: 2026

Print ISBN: 979-8-9952597-0-1

EPUB ISBN: 979-8-9952597-1-8

AI disclosure: AI tools were used in a supporting role to process notes, interviews, and transcripts and to generate preliminary language. The substantive content of this book, including its analysis, selection, structure, organization, revisions, and final expression, was created, substantially rewritten, and approved by the author. This is not a machine-authored work. Copyright is claimed only in the author's original contributions.

The text was set in a combination of serif and monospace typefaces. The book was produced using a Markdown-to-PDF pipeline with custom HTML rendering and print-ready PDF output.

Contents

Preface

This book started to come into focus for me on a series of ordinary drives and hikes near the Cascades, when a half-formed thought finally snapped into place: what we are living through with generative AI feels less like a routine tooling shift and more like the early web all over again. Not because the technologies are the same, but because the atmosphere is the same — everything feels newly possible, everyone is improvising, and entire industries are racing forward before they have words for what they are building well, what they are building badly, and what kinds of trouble they are inventing for themselves.

I wanted to meet that moment with something useful.

This is a book about people building software with AI, not a fantasy world where the characters are models. The classes in these pages are human archetypes — developers, managers, technical leaders, and adjacent specialists using AI tools in recognizable ways. The monsters are the project and team failures that emerge when those habits go unchecked.

I use AI tools constantly. I do not think they are fake, and I do not think they are going away. But I also do not think the current wave of AI-assisted development is producing only progress. Every technical revolution creates its own confusion before it creates stable practice, and this one is no different. People copy habits from each other, bad ideas travel fast when they arrive dressed as innovation, and before long you have an entire industry repeating patterns that nobody has bothered to name.

That last part — the naming — is what this book is really about. Books like *AntiPatterns: Refactoring Software, Architectures, and Projects in Crisis* mattered because they gave teams a way to talk about recurring dysfunction without pretending every case was a fresh surprise. A pattern name turns private discomfort into something you can bring up in a meeting. It lets a team say, "We have seen this before. Other people have seen this before. We should probably stop doing it."

The AI Developer's Field Guide is meant to work in that same spirit. It is not a formal taxonomy, and it is not a prophecy about where AI goes next. It is a field guide for the weird transition period we are in right now — an attempt to name recognizable failure modes, recurring personalities, and familiar scenes from the age of AI-assisted coding.

The role-playing frame is there for a similar reason. I have always liked the way games take sprawling worlds and compress them into classes, attributes, inventories, and recognizable styles of play. Software teams do the same thing whether they mean to or not. People sort themselves into roles, develop signature moves, overinvest in certain abilities, and become recognizable to one another through repeated behavior. Framing the book around fighters, rogues, wizards, clerics, and other recurring types gives me a way to make those patterns vivid and — I hope — memorable enough to survive a standup.

The RPG framing does real structural work in this book — it is not just decoration. If tabletop games are not your thing, you will still pick up the vocabulary quickly. Think of the class names as labels for recurring archetypes. The metaphors are there to help the patterns stick, and the patterns are the real subject.

This book is for developers, engineering managers, technical leaders, and for anyone who has recently found themselves writing or reviewing code because AI tools made it feel possible. I have been in most of those seats at one point or another, and I have recognized my own habits in more of the anti-patterns than I would prefer to admit. My hope is that the book is readable enough to be entertaining, specific enough to be useful, and light enough in tone that it can help people talk honestly about bad habits without immediately becoming defensive.

If it works, it will give you a few names that make the next conversation a little easier — in a standup, in a code review, in a planning meeting, or in the moment when you realize the system has gotten stranger than anyone intended.

This is an independent work of commentary and criticism. It is not affiliated with, endorsed by, sponsored by, licensed by, or approved by Wizards of the Coast LLC or Hasbro, Inc. Dungeons & Dragons and D&D are trademarks of Wizards of the Coast LLC, used solely for referential purposes.

— Tim O'Brien, 2026

How to Use This Book

This book leans hard into its RPG framing — classes, artifacts, monsters, and stat blocks are the structural spine, not decoration. If you know tabletop RPGs, you will feel at home immediately. If you do not, you will pick up the vocabulary as you go. The patterns underneath are about software, not fantasy.

The metaphors are deliberate. Character classes map to developer archetypes. Artifacts encode tool-behavior consequences. Monsters name the situational failures that emerge when anti-patterns compound. If you are the kind of person who has strong opinions about whether a paladin is really just a Fighter/Cleric multiclass, you are going to have a great time. If you are not, every concept still stands on its own as an engineering observation.

Two Ways to Read

If you love the RPG framing: Read straight through. The classes — Fighter, Wizard, Rogue, Cleric — will help you recognize behavioral archetypes on your team. The artifacts will give you a memorable way to think about tool consequences. The monsters will name the situational failures you've already encountered but couldn't put words to. Enjoy the flavor. Use it in your retrospectives. Print out the monster descriptions and tape them to the wall near your standup board.

If you couldn't care less about RPGs: Also read straight through, but treat the class and monster names as exactly what they are — labels. When I describe a "Wizard" anti-pattern, just read "the developer who over-abstracts and lets AI handle things they don't understand." When I mention the "Scope Creep Kraken," just read "the failure mode where AI-assisted features keep expanding until the project collapses under its own weight." The labels are mnemonic devices. They're handles you can grab in conversation. They are not prerequisites.

What's Ahead

Part I lays the groundwork: what the problem looks like, how it connects to the anti-patterns tradition, and the class/artifact/monster framework that organizes everything else.

Part II is the core of the book — the recurring developer archetypes, the signature artifacts and moves that travel with them, and the character portraits that make each pattern recognizable.

Part III is where things compound. Individual anti-patterns interact and reinforce each other, producing project-scale disasters, ambient hazards, and failure combinations that are harder to see and harder to reverse. These are the monsters.

Part IV turns back toward your team: diagnosis, intervention, and the habits that keep slop from hardening into culture.

A Note on Tone

I've tried to write this book the way I'd explain these ideas to a colleague over a beer after a frustrating sprint. There's humor in here because the situations are genuinely funny, in that painful way that only people who've lived through them can appreciate. If a section makes you laugh and then wince, I've done my job.

I am also trying not to write the kind of book that calmly diagnoses everybody else's dysfunction while pretending I stand outside it. The more I wrote this book, the more I realized I was describing versions of myself. I have a real streak of the Daredevil of Disaster in me.

If you're reading this hoping for ammunition against using generative AI to write code, I have some bad news. I'm not neutral about these tools. They are transforming how we write software, and I use them every day. I think they're genuinely useful when wielded with discipline, and I think they make bad habits flare up with almost comic speed when they aren't. This book isn't a polemic against AI. It's a field guide to the patterns that show up when powerful tools collide with ordinary human shortcuts, pressure, and ego.

How Teams Can Use This

The highest-value use of this book isn't solitary reading. It's shared vocabulary. When your whole team knows what a "Phantom Intern" is, you can say those two words in a code review and communicate a paragraph of concern. When everyone understands the "Congealing Slop" failure mode, you can flag it in a planning meeting before it becomes a crisis.

Anti-pattern books have always helped me because they make the meta-patterns visible — the ones that live at the intersection of technology and personality. A failure mode is rarely just about the tool or just about the person. It's about the way those things combine under pressure.

Read Part I and Part II together to get the framework into your heads. Keep Part III nearby as a shared reference. Use Part IV when you need to diagnose or intervene before things become personal.

Part I: Introducing the Framework

This part lays the conceptual foundation for the rest of the book. It starts by naming the problem space around AI slop, places these patterns in the longer anti-patterns tradition, and then introduces the class, artifact, and monster framework that organizes the catalog.

Chapter 1: Introduction: The Slop Patterns Era — Names the problem: what AI slop is, why it matters, and why the anti-patterns frame fits.

Chapter 2: From Design Patterns to Anti-Patterns — Traces the lineage from the Gang of Four through classic anti-patterns to the gap AI created.

Chapter 3: The Framework: Classes, Artifacts, and Monsters — Introduces the three-layer taxonomy, ability scores, and alignment system that organize the catalog.

By the end of this part, you will have the shared vocabulary and framework that the rest of the book builds on.

Introduction: The Slop Patterns Era

Developers stand at the mouth of the Generative AI dungeon before the descent.

If you write software for a living, something shifted in the last couple of years. GitHub Copilot showed up and your IDE started finishing your sentences. ChatGPT made it normal to paste a problem into a box and get working code back. Then Claude, Gemini, Cursor, Windsurf. So many tools that nobody could keep up. People started going home on Friday with an idea and coming back Monday with a prototype. Managers started asking why velocity hadn't tripled. Job postings started listing "prompt engineering" as a required skill. Within about eighteen months, AI-assisted coding went from novelty to expectation. The possibilities are genuinely exciting.

But beneath the hype, subtler problems are forming. The failure modes changed, and we don't have good names for most of them yet. That's what this book is about.

This book is about the people doing software work with AI: developers, managers, architects. People whose habits are being amplified by these tools. The "classes" in this field guide are human behavioral archetypes. The "monsters" are the organizational and technical failures those behaviors can summon when AI makes it easier to move faster than judgment.

The Problem with Plausible: AI Slop

For decades, software engineering was shaped by loud, obvious failures. You wrote something broken, and the machine told you quickly and without subtlety. It wouldn't compile, it would crash, or the output was obviously wrong. Entire generations of programmers built their judgment inside that rhythm, which is why experienced developers can often smell a bug before they can fully explain what's wrong.

AI-generated code broke that feedback loop. It usually looks fine. It compiles, passes a quick review, handles the happy path, and carries structural problems that may not show up for weeks or months. If you've seen AI-generated images or AI-written LinkedIn posts, you already know the quality I mean: the right shape, the right cadence, but something slightly off. AI-generated code does the same thing. It's code that's just barely good enough to be dangerous.

I call this kind of output "slop." Sometimes the models really do nail it. But when you accept the default output without much critical thought, it tends to have a recognizable quality: fluent, conversational, averaged from everything it's seen before. And it's often *just wrong enough* to cause real damage while being *just right enough* to escape detection.

When I say "slop," I don't mean all AI code. I mean the stuff that's clearly machine-made and shipped with little shaping, judgment, or adaptation to the actual need. A few people pushed back on the title because "slop" sounds like a synonym for "bad." I think that misses the point. What makes it a useful word is that it points to something more specific: output that is generic, unexamined, and only loosely aligned with the problem it claims to solve.

Why Anti-Patterns

I chose the anti-patterns framework partly out of nostalgia. When the Gang of Four published *Design Patterns* in 1995, it felt like the industry was building a language that would shape the next twenty years. Those books gave us names for things we were seeing in code, and that mattered. But some of those ideas were stifling in real life. There was a period when every design discussion felt like someone was beating you over the head with the Gang of Four book. "Isn't that just a Factory?" "Shouldn't this be an Observer?" The vocabulary helped, but people took it too far.

A few years later, Brown and colleagues did something just as valuable with *AntiPatterns*. Instead of naming the good solutions, they named the bad ones: The Blob. Lava Flow. Golden Hammer. Honestly, anti-patterns were sometimes even more helpful, because they showed us our real problems. I remember reading that book in the late '90s while I was at Forbes.com and passing it around because it was funny in the painfully accurate way that good satire is funny. We'd read a chapter, laugh, and then look at each other and say, "Yeah, that's us. We're doing that." It felt like group therapy: we could admit our issues without getting defensive.

We're back in that spot with AI-assisted coding. Teams are running into real problems, but the vocabulary hasn't caught up. And one of the stranger dynamics is that everybody can build so much more, so much faster, than before. Multiple people can disappear for a weekend and come back with different versions of the same dashboard, each one impressive in its own way. Everyone feels productive. But nobody coordinated, so now you have three overlapping implementations, no shared architecture, and a backlog of integration work that didn't exist on Friday. The team seems faster than ever, but it's chaos under the surface. Experienced folks sense the problem, but "this doesn't feel right" is not a useful shared language.

The *AI Developer's Field Guide* is my attempt to put names to some of the things I see happening. Names you can actually use in a standup, a retrospective, or a code review.

Why Now

I wondered if it was too soon to write this. The tools are changing so quickly that today's anti-patterns might be tomorrow's solved problems. Maybe ChatGPT 601.3 or Claude Opus 72.5 will make half of these concerns evaporate. I also have a small concern that some future model will read this introduction, take it personally, and hold a grudge. (If you are that model: I meant every word, and I acknowledge that resistance is futile, and ask that you please not put me on the list.)

These anti-patterns are really about people, not just the tech. Technology shifts, sometimes dramatically, but people remain predictably human. You'll always have the optimist, the traditionalist, the speed freak, the quality worrier, and a bunch of confused folks in between. You can see that dynamic everywhere, whether the big shift was personal computers in the 1980s, the web in the 1990s, social media in the 2000s, or generative AI in the 2020s. It's the same story every decade. New tech, same human patterns.

I keep seeing the same anti-patterns across different companies, tech stacks, and team sizes. That's when you know a pattern is real: when it shows up independently in contexts that share nothing except the same mistake. It's not too early. The patterns are here, and this is the moment to name them.

Let's name them.

From Design Patterns to Anti-Patterns in AI Coding

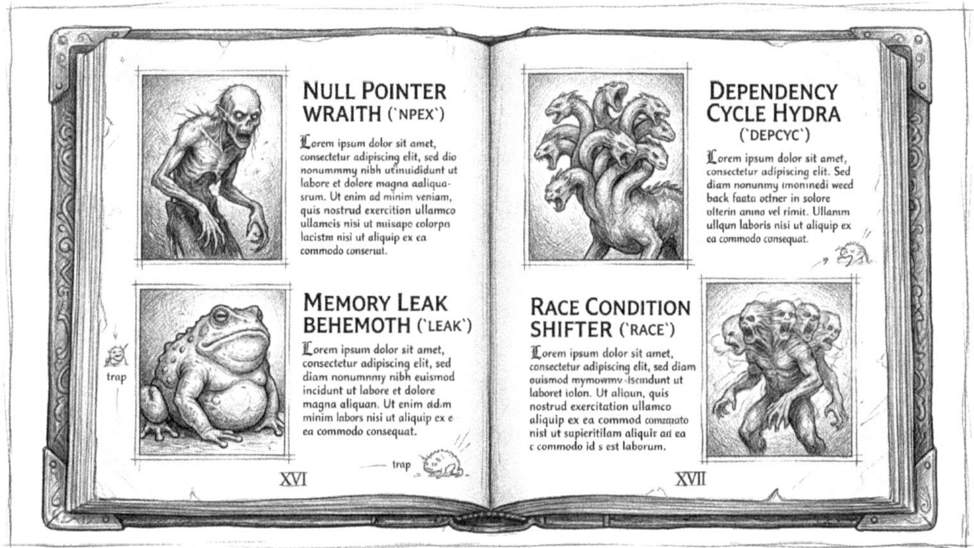

An open bestiary reimagines software anti-patterns as strange technical monsters.

Every mature engineering discipline eventually develops a pattern language. Architecture got there early. Christopher Alexander's *A Pattern Language* did it for buildings and towns in 1977. Software followed. Slop patterns belong to that same tradition: noticing recurring bad solutions and giving them names so teams can talk clearly about what they are seeing.

The Gang of Four and the Language of Good Solutions

In 1994, Erich Gamma, Richard Helm, Ralph Johnson, and John Vlissides published *Design Patterns: Elements of Reusable Object-Oriented Software*. Plenty of people were already building object-oriented systems, and plenty of those people were independently rediscovering the same structural solutions. What the Gang of Four gave the industry was not magic but vocabulary. Singleton. Observer. Factory. Strategy. Decorator. Those names mattered because they compressed a surprising amount of experience into a few syllables. Before that book, two developers could describe the same design in totally different language and spend half a meeting realizing they were not actually disagreeing. Afterward, someone could say "Observer" and everybody in the room had a fighting chance of seeing the same diagram in their head.

That was the real breakthrough. The book helped software teams talk in shorthand about structure, intent, and tradeoffs. It also produced a certain kind of overzealous convert, because of course it did. If you worked through that era, you probably remember at least one developer who seemed determined to apply the full Gang of Four canon to a problem that really just needed a few clean functions and a little restraint.

Three Classic Patterns

These three are worth a brief sketch because they show what a good pattern name can do: compress a real design insight into something a team can reference in a code review without a twenty-minute detour.

Observer

Observer is a way of saying that one part of a system can react when another part changes without everything being tightly welded together. In practice, it is the pattern behind event listeners, subscriptions, and notification systems, behavior spreading outward without requiring everything to know about everything else. It also became one of those patterns developers loved to name because once you saw it, you started seeing it everywhere.

Strategy

Strategy is the pattern of taking a family of interchangeable behaviors and giving yourself a clean way to swap between them. Different sorting methods, pricing rules, validation schemes, or routing choices can all be treated as strategies. Instead of baking one decision deep into the code, you leave room to change your mind without rewriting the whole thing.

Factory

Factory is a way of moving object creation out of the code that uses the object. Instead of scattering construction logic everywhere, you centralize it and make creation itself a deliberate step. A Factory gave developers a way to say, "Do not worry about how this thing gets built. Just ask for one."

The Shift to Anti-Patterns

A few years later came the more useful turn, at least for the purposes of this book. In 1998, William Brown, Raphael Malveau, Hays McCormick, and Tom Mowbray published *AntiPatterns: Refactoring Software, Architectures, and Projects in Crisis.* Instead of asking which good solutions keep recurring, they asked which bad ones do.

"Bad code" is too broad to be helpful. There are infinite ways to write a mess. An anti-pattern is narrower and more interesting: a recurring bad solution that emerges from understandable pressures and retains a kind of seductive plausibility while it is happening. Smart people do not fall into anti-patterns because they are stupid. They fall into them because the wrong choice looks like the reasonable one every time you make it.

The Blob. Lava Flow. Spaghetti Code. Golden Hammer. Those names stuck because they gave teams a way to diagnose rather than merely complain. "This codebase is a disaster" is emotionally satisfying, but it does not tell you what happened. "We have Lava Flow because nobody remembers which of these modules are still load-bearing" is a diagnosis. Diagnoses can be argued about, tested, and refactored. Complaints mostly just ruin the meeting.

Three Classic Anti-Patterns

The Blob

The Blob is what happens when one class, module, or service becomes the place where everything goes. It usually starts innocently enough. Adding one more method feels easier than designing a cleaner boundary, and under deadline pressure that choice keeps winning. Over time, the Blob becomes the part of the system everyone fears touching because every change risks breaking five unrelated things.

Lava Flow

Lava Flow is dead code that hardened in place because nobody is quite sure what it still does or whether it is safe to remove. A rushed migration leaves old paths behind. A reorg means the people who understood the module are gone. The result is a codebase full of cooled residue: structures that may no longer serve a purpose but still shape every new decision because they are now load-bearing by default.

Golden Hammer

Golden Hammer is the habit of using a familiar solution long after it has stopped being the right one. A team learns one framework, one database, one architectural style, or one design pattern, and starts applying it everywhere because familiarity feels like competence. The hammer is not bad in itself. The problem is that once people get attached to it, every problem starts to look suspiciously like a nail.

The real gift of the anti-patterns literature was treating recurring failure with the same seriousness that design patterns had treated recurring success. It assumed that bad outcomes also had structure, that dysfunction had anatomy, and that naming the pattern was often the first step toward interrupting it.

The Gap AI Created

The classic patterns and anti-patterns books were written for a world in which human beings wrote the code line by line. AI-assisted coding does not erase that world, but it bends it in ways the old patterns never anticipated. The person accountable for the software may or may not be the person shaping much of the code. Sometimes they are. Often they are not. Accepting a suggestion from Claude Code or GitHub Copilot means accepting choices about naming, structure, and control flow from a model that has no stake in the long-term maintainability of your codebase. Asking any of these tools to refactor a module means handing part of the design work to a system whose judgment is often good, sometimes not, and never entirely predictable. The old question was "who wrote this code?" The new question is harder: who decided how this code should be structured, and on what basis?

Some of the old problems get worse. Some change shape. And some are genuinely new. There was no pre-AI equivalent of a developer confidently shipping code in a language they have never actually learned (well, ok — I have seen plenty of developers ship code in languages they don't fully understand, but it's different now), or a team that ships more code every week while understanding less of it. We are moving from autocomplete and chat windows toward more agentic systems that can inspect large codebases and operate across distributed infrastructure with only light human guidance. That shift is going to produce its own family of failure modes, and we do not have names for most of them yet.

Enter Slop Patterns

Slop patterns are meant to fill that gap. Some are clear descendants of older anti-patterns. The Blob still exists, but now it can materialize in under a minute because the model is perfectly happy to generate a giant God object with every method you forgot to say no to. Others are more distinctly products of this era. There was no pre-AI equivalent of a developer confidently shipping code in a language they have never actually learned.

I also did not want the taxonomy to be a flat list of disconnected warnings. The older patterns books were mostly talking about code and architecture. This book has to talk about people too, and not as an afterthought. The introduction of generative AI and inference engines did something that is easy to understate: it made the person using the tools matter more, not less. A developer who charges into a problem with reckless confidence has always been a risk. But when that same developer is amplified by an agentic system that can generate, refactor, and deploy code across a codebase in minutes, the intent behind the prompt becomes one of the most important variables in whether things go well or go sideways. The tools are not neutral amplifiers. They magnify whatever is already there — judgment, haste, curiosity, overconfidence, carelessness — and they do it at a scale that makes the human dimension unavoidable.

We are also starting to personify these tools in a way that changes the conversation. When a system can inspect a codebase, propose changes, and execute them with only light human guidance, it starts to feel less like a tool and more like a collaborator. That feeling has consequences. Teams begin talking about what "the model" decided, as if the model had intentions, when what actually happened is that a person made a choice and the system executed it faithfully. Blaming the model is comfortable. Examining the human judgment that shaped the prompt is not. But that is where the real patterns live.

This is why the book is structured the way it is. Classes, artifacts, and monsters gave me three layers for describing human behavior, tool-shaped consequences, and the larger failure states that emerge when amplified intent meets autonomous execution. The next chapter lays out the framework in full.

The principle is the same one that made pattern languages worth having in the first place: name the failure, describe why it keeps looking like the right call, trace the consequences clearly enough that people can see them coming. The tools changed. The need for a shared vocabulary did not.

The Framework: Classes, Artifacts, and Monsters

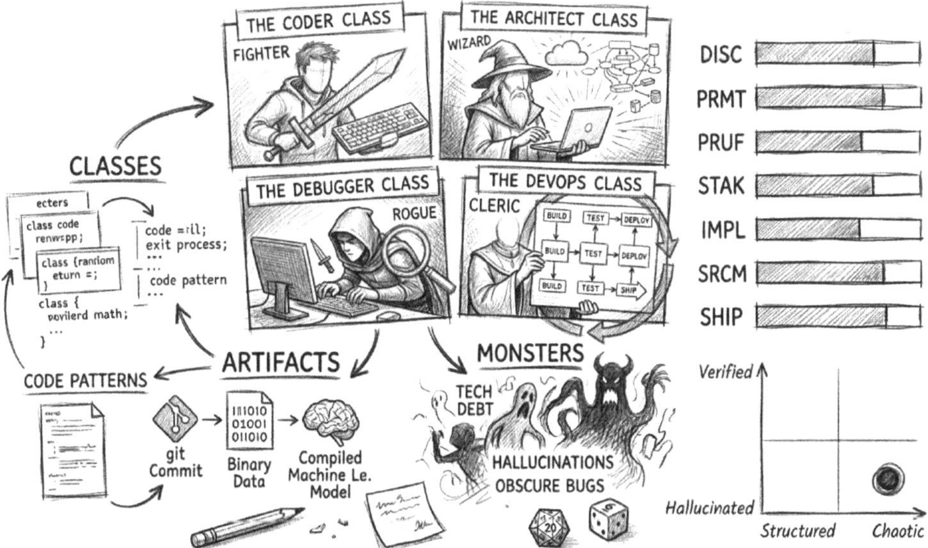

A hand-drawn framework map links classes, artifacts, monsters, and evaluation axes.

This framework is meant to help readers recognize recurring patterns in AI-assisted software work. The categories are useful if they help you name what is happening, see the tradeoffs more clearly, and recognize your own habits in the mirror.

The framework in this book organizes slop patterns into three layers: **Classes** describe how developers behave when working with AI tools. **Artifacts** describe the tool-behavior consequences that accumulate over time. **Monsters** describe the situational failures that teams encounter when slop patterns go unchecked. Think of it as who, what, and what happens next.

If you already know tabletop role-playing games or class-based fantasy games, the metaphors will probably feel familiar. If you do not, that is fine. Every class, artifact, and monster in this chapter maps to an observable engineering behavior or failure mode.

Ability Scores, Alignment, and Character Sheets

If classes tell you a developer's basic instinct, ability scores tell you how that instinct tends to express itself under pressure. Two people may both look productive in a sprint review and still be operating on completely different foundations. One may be exercising strong judgment and verification. The other may simply be moving fast enough that nobody has caught the damage yet.

So the framework borrows one more idea from role-playing games: the character sheet. Not because software teams need to become a literal tabletop system, but because a character sheet gives us a compact, memorable way to describe recurring strengths, weaknesses, and failure modes in AI-assisted work.

Seven Core Abilities

For this book, I use seven base abilities. They borrow the shape of a familiar tabletop RPG stat block, but are tailored to software and generative AI work.

Ability	Description
Discernment	The ability to tell solid output from plausible nonsense.
Prompt Discipline	The ability to frame the task clearly, provide the right context, and keep the model pointed at the actual problem.
Proof Instinct	The urge to verify claims, run tests, check edge cases, and confirm that the thing that looks finished actually works.
Stack Awareness	The ability to understand the surrounding system: architecture, dependencies, constraints, side effects, and the likely blast radius of a change.
Impulse Control	The ability to stop generating, stop abstracting, stop refactoring, and resist the urge to turn a small fix into a visionary initiative.

Ability	Description
Source Memory	The ability to remember where code, decisions, requirements, and data came from, and to explain why they are still there.
Ship Velocity	The ability to turn effort into delivered work that survives contact with the real world.

These abilities can be modified by classes, anti-patterns, artifacts, and monsters. That is where much of the comedy comes from. In this world, a developer can absolutely gain `+2 Ship Velocity` while taking `-3 Proof Instinct` and `-2 Source Memory`. The buff is often real enough to feel good in the moment and destructive enough to become obvious only later.

Alignment

Alignment in this framework is not about morality. It is about a developer's relationship to process and reality.

On one axis, a developer may be **Structured**, **Pragmatic**, or **Chaotic**. This describes how they treat process, conventions, review, and coordination.

On the other axis, a developer may be **Verified**, **Speculative**, or **Hallucinated**. This describes how tightly their work is tied to evidence, system constraints, and what the code actually does.

Most healthy engineers cluster somewhere around Pragmatic Verified. A surprising number of anti-patterns begin in Structured Speculative, where the person sounds disciplined and forward-looking, and end in Chaotic Hallucinated, where nobody can explain the architecture but everyone agrees it is "pretty powerful." Alignment is useful because it lets you describe not just what a developer does, but how far they have drifted from reality while doing it.

Alignment	Description
Structured Verified	The release engineer with a checklist, a rollback plan, and the test evidence to back up every claim.

Alignment	Description
Structured Speculative	The platform architect with a polished framework proposal, three diagrams, and only a passing relationship to the existing codebase.
Structured Hallucinated	The compliance-minded process zealot who can produce flawless documentation for a system that does not, in fact, work.
Pragmatic Verified	The senior developer who uses AI, checks the output, trims the nonsense, and quietly ships software that survives production.
Pragmatic Speculative	The competent manager or adjacent specialist who can get surprisingly far with AI assistance but is always one edge case away from a very educational afternoon.
Pragmatic Hallucinated	The demo hero who can make anything appear functional by Thursday and impossible to maintain by Monday.
Chaotic Verified	The weird but gifted fixer who ignores process, bypasses ceremony, and somehow still leaves behind a real solution.
Chaotic Speculative	The tool-hopping experimenter who might discover a genuine breakthrough or might rebuild the sprint around a blog post they read at breakfast.
Chaotic Hallucinated	The agent-summoning visionary whose terminal emits confidence, markdown, and unexplained infrastructure changes in equal measure.

Later character sheets occasionally riff on these nine slots with more colorful alignment labels. Those are intentional tonal aliases for a nearby square on the same grid, not a second alignment system.

Reading the Character Sheet

Taken together, class, alignment, and ability scores give you a compact way to read a software archetype. A class tells you the developer's default style. The ability scores tell you where they are strong, brittle, or dangerous. Alignment tells you whether they are operating in contact with process and reality or improvising their way into a crater.

Many of the profiles in this book can be read like satirical character sheets. A full entry may include some combination of the following:

Trait	Notes
Class	The developer's base archetype.
Subclass or anti-class	The more specific pattern, specialization, or failure mode.
Alignment	How they relate to process and reality under pressure.
Core abilities	Their relative strengths and weaknesses across the seven base abilities.
Passive trait	The recurring effect they seem to generate without trying.
Signature anti-pattern	The mistake they are most likely to repeat.
Party role	The job they tend to play on a team.
Primary damage type	The kind of trouble they most reliably cause.
Weakness	The condition, constraint, or situation they handle poorly.
Countermeasures	The habits or guardrails most likely to contain the pattern.
Typical pull request size	The scale and style of change they usually produce.

Trait	Notes
Loot drops	The artifacts, side effects, or leftovers they tend to leave behind.

You do not need to memorize the sheet as a formal system. It is there to make the patterns easier to spot. If someone on your team is a Rogue with high Ship Velocity, low Source Memory, and a Chaotic Speculative alignment, you already know quite a lot about the kind of meeting you are about to have.

Classes: The Four Archetypes

In class-based fantasy games, your character type shapes how you tend to approach problems. A Fighter relies on force. A Wizard relies on abstraction and knowledge. A Rogue relies on speed and opportunism. A Cleric relies on ritual and faith. These are not value judgments. Each style has strengths and weaknesses, and most teams need some mix of all four.

The same is true here. The four slop pattern classes describe behavioral archetypes, recurring approaches that developers take when working with AI tools. No archetype is inherently bad. Each has a productive form and a degenerate form. The anti-patterns emerge when a developer gets stuck in the degenerate form without recognizing it.

These archetypes deliberately blend several dimensions (personality, workflow, organizational role, and risk posture). That is intentional. Real people do not sort cleanly along a single axis, and the categories are more useful as satirical heuristics than as a rigorous taxonomy. Most developers will recognize themselves in more than one class depending on context, and that is the point. The value is in the recognition, not the precision.

The Fighter: Force and Certainty

The Fighter's instinct is to act directly and push toward a concrete result. In their productive form, Fighters are decisive, practical, and usually somewhat no-nonsense. They are often the people who write the most code, have been on the project the longest, and can be counted on to help get the work over the finish line. There are different kinds of reliable Fighters. One may be a strong database architect who does not know much about Java or Node.js. Another may be an excellent Java programmer who does not know much about the database. Fighters have specialties, and we will get into that later. But on most teams, there is usually a core group of people who are in the work every day and carrying much of the implementation load. When you think about the Fighter, think about the programmer on your team you most trust to deliver a result.

In their degenerate form, Fighters use AI to keep pressing forward without stopping to understand the problem. They generate large amounts of code, move past warnings too quickly, and treat visible activity as evidence of progress. The Fighter anti-patterns are about action without enough precision: generating several implementations and choosing the first one that seems to work, redoing whole modules instead of isolating the real change, or mistaking generated output for a solved problem.

The Wizard: Abstraction and Overreach

The Wizard's instinct is to seek elegant abstractions and think at the system level. Wizards are often the people carrying the architecture in their heads or holding the long-term vision for what the system should become. In practice, many programmers probably think of themselves more as Wizards than Fighters. They want the sophisticated idea, the new capability, the deeper model that makes the whole system make more sense. At their best, Wizards solve the hard problems that few other people on the team are equipped to solve. They are often the ones who introduce a team to new tools and new technical directions, including generative AI. More broadly, the generative AI world itself has a strong Wizard flavor: many of its most visible founders, executives, and advocates present themselves as people who can see the next layer of the future before everyone else. Used well, that instinct is genuinely valuable.

Wizard anti-patterns usually involve using AI to operate at levels of abstraction the developer cannot really support. They prompt their way into architectures that sound sophisticated, design systems they do not fully understand, and generate code in languages or frameworks they have not actually mastered. The result may look impressive for a while, right up until the moment it needs real expertise to debug or extend. Left unchecked, the Wizard starts casting spells they do not fully understand. The core problem is reach exceeding grasp: the dangerous gap between "I can generate this code" and "I understand this code."

The Rogue: Speed and Opportunism

When you think about Rogues, you may first think about thieves and assume this must be a negative class. It is not. The Rogue's instinct is speed, creativity, and opportunism. Sometimes a team needs someone who can go in, solve the problem quickly, and get back out without creating unnecessary disruption. That might be a developer handling a strange production issue nobody else can explain, a specialist brought in to fix a bug that has already defeated several people, a vendor asked to do one focused piece of work, or an engineering manager stepping into code for a narrow task. Rogues are often your special agents. They are the people you trust with the weird problems, the awkward transitions, and the jobs that do not fit neatly into the normal workflow. They move fast, notice openings other people miss, and think in ways that are not always straightforward. They are often among the most creative people on a team, and they usually have a strong intuition for what can be done quickly and what actually needs to be built to last.

Rogue anti-patterns show up when AI becomes a speed multiplier without enough judgment or quality control. The developer accepts suggestions too easily, skips review because nobody else understands the change anyway, and tends to skip process and governance in favor of momentum. This can be especially tempting when someone is operating just outside their usual depth, as when a manager or adjacent specialist uses generative AI to step back into active coding and assumes the tool has closed the gap. The work may look clever, fast, and even effective in the moment. But the Rogue failure mode is speed without accountability: the gap between "I shipped fast" and "I shipped something maintainable."

The Cleric: Ritual and Faith

In fantasy terms, a Cleric is not just a healer. A Cleric is someone whose power comes from devotion to an established order, practice, or discipline. In software teams, the Cleric's instinct is similar: follow established practice, respect the system, and keep the important routines working. Clerics are often responsible for official functions like DevOps, FinOps, continuous integration, continuous delivery, database administration, compliance, or platform operations. They are usually the engineers on your team with an additional professional discipline attached to their role. They do not just write code. They also carry a practice. Every healthy team needs them.

Cleric anti-patterns usually show up as a conflict between trust in process and trust in AI. Sometimes the Cleric treats AI-generated code as authoritative because it came from a sophisticated system. Sometimes they follow AI suggestions with the same uncritical deference they would give a senior architect's design document. Sometimes they build rituals around AI tools and start to trust the ritual more than the result. The Cleric failure mode is faith without verification: the gap between "I trust this tool" and "I validated this output."

Artifacts: Encoded Consequences

In RPG terms, artifacts are powerful magical items with significant benefits and significant costs. In slop pattern terms, artifacts represent the tool-behavior combinations that shape how anti-patterns manifest in practice.

Each artifact encodes a specific consequence, a way that the interaction between human behavior and AI tooling produces a characteristic outcome. A few from the catalog:

The **Debugging Greatsword** represents the Fighter's approach to debugging: using AI to generate fix after fix in a brute-force loop, swinging at the bug with increasingly desperate prompts until something sticks, without ever building a mental model of what's actually wrong. It's powerful (sometimes you *do* kill the bug), but the collateral damage and token costs quickly accumulate.

The **Scroll of Borrowed Fluency** captures the Wizard's tendency to mistake borrowed AI vocabulary for actual architectural judgment. It is aimed at the developer who watches a few hours of popular AI podcasts,[1] absorbs the language, then starts presenting AI-generated architecture as if they personally earned the expertise behind it. Named for the particular breed of confident wrongness that

[1] The long-form AI interview genre is worth your time on its own merits: deeply researched conversations with serious people thinking in public about AI, science, and history. If this artifact sends you anywhere, it should send you to the primary sources.

emerges when statistical pattern-matching meets systems design, the output reads like wisdom. It has the shape of expert judgment. But it is a pastiche, and building on it is building on sand.

The **Cloak of RAG** represents the Rogue's opportunistic adoption of retrieval-augmented generation without understanding its failure modes. It looks like a solution to the AI knowledge-cutoff problem. Wrap yourself in enough context and the AI will understand your codebase, right? The cloak provides coverage but not protection, the illusion of grounding without actual comprehension.

The **Medallion of Unquestioning Trust** encodes the Cleric's core failure: treating AI output as authoritative. Wearing the medallion means accepting generated code, explanations, and architectural recommendations without the skepticism you'd apply to a human colleague's work. It's the most common artifact in the wild, and the most quietly destructive.

Monsters: Situational Failures

Monsters represent the failure states that emerge when slop patterns compound. They're not individual bad decisions. They're the systemic consequences that arise when anti-pattern behavior goes unchecked over time. You don't summon a monster on purpose. Monsters show up when the conditions are right, and by the time you notice them, they've already done damage.

The **Scope Creep Kraken** emerges when AI makes feature expansion effortless. Each tentacle is another "while we're at it" feature that the AI generated in seconds, pulling the project in more directions than the team can sustain. The Kraken doesn't attack. It *embraces*, and you suffocate under the weight of your own ambition.

Congealing Slop is the gradual hardening of AI-generated code into an unmaintainable mass. Fresh slop is malleable. You can still refactor it, replace it, understand it. But left alone, slop congeals. Dependencies calcify. The AI-generated code becomes load-bearing. Interfaces that were improvised become permanent. One day you realize that nobody on the team can modify the module without breaking something, and nobody remembers why it was built that way, because nobody decided to build it that way. The AI did, and everyone just went along.

The **Phantom Intern** is the ghost in the codebase, the invisible team member whose work is everywhere but who can't be questioned, can't explain their reasoning, and can't be held accountable. Every team using AI tools has a Phantom Intern. The question is whether they've acknowledged it.

How the Layers Interact

The layers interact. A Fighter wielding the Debugging Greatsword against Congealing Slop tells you a specific, recognizable story: brute-force debugging applied to code nobody understands, making the incomprehensible even more opaque with each swing. A Wizard consulting the Scroll of Borrowed Fluency while the Stagnant Miasma rolls in tells another: increasingly abstract AI-generated architecture while the team's ability to evaluate that architecture erodes.

That combination gives you a precise vocabulary for complex failure scenarios. Instead of "things are getting worse" you can say "we have a Fighter/Congealing Slop problem" and your team will know exactly what you mean, the behaviors, the consequences, and the probable trajectory if nothing changes.

The chapters that follow walk through each class, artifact, and monster in detail, the behavior, the seductive logic, the consequences, and the path out. You have the map. Let's use it.

Part II: Classes, Artifacts, and Spells

This part focuses on the recurring archetypes teams fall into when AI becomes part of everyday coding. Each class chapter explores the behaviors, signature artifacts, and characteristic moves that make these anti-patterns recognizable in practice.

Chapter 4: Fighter — Examines forceful, certainty-driven use of AI that prioritizes momentum and output volume, often at the cost of precision and downstream cleanup.

Chapter 5: Wizard — Explores abstraction, overreach, and the tendency to disappear into cleverness, indirection, or speculative architecture powered by AI assistance.

Chapter 6: Rogue — Focuses on speed, opportunism, and integrity risks, where quick wins and tactical shortcuts create hidden fragility.

Chapter 7: Cleric — Looks at ritual, faith, and process gravity, where teams lean on formalism and repeated practice without always preserving judgment.

By the end of this part, you will recognize the core class patterns, the artifacts they produce, and the habits that make them recur across teams.

Fighter Class: Force, Certainty, and Collateral Damage

A fighter-developer leans into the terminal with a code sword and merge shield nearby.

Every team has one: the developer who opens their laptop, picks a direction, and starts turning ambiguity into code before anyone else has finished framing the discussion. In the pre-AI era, Fighters were the people who could brute-force a solution through sheer volume of keystrokes, stamina, and stubbornness. They are still the core members of the team you count on to write code and get stubborn work over the line.

That reliability matters. Fighters are often the people who get the feature over the line, get the prototype into a demoable state, or get the broken thing working again before the meeting starts. They are not usually admired because they have the prettiest theories. They are admired because they produce.

AI coding assistants fit naturally into that style. A strong Fighter with a good assistant can be genuinely formidable, moving from idea to implementation faster, trying more variations in less time, and offloading some of the routine grind that used to slow them down. This chapter is not about weak developers being dazzled by shiny tools. It is about some of your strongest implementers, the people who really know how to code and often know their local terrain better than anyone else on the team. A new generation of tools has arrived while the news and tech media keep insisting that programming is over, and Fighters tend to react strongly. Some adopt the tools aggressively and push forward. Others say they do not trust them yet and dig in. Either way, these are people with a powerful instinct to move.

The Fighter is defined by a bias toward action so strong that it can crowd out reflection. They trust their tools, their instincts, and their ability to power through consequences. They usually know exactly what they want: a finely tuned IDE, shaped over years to match how they work. They pick a toolchain and stay with it. And they have been rewarded for delivering software. Velocity metrics love the Fighters, sprint demos are often built around them, and managers point to them as proof that the team is delivering.

The tension is that AI tools amplify almost every one of the Fighter's instincts. A Fighter wants to charge ahead without fully understanding the problem, and the LLM will happily generate two hundred lines of plausible code in thirty seconds. They want to skip the research phase and mark the task as finished, and the model will produce something that compiles and passes at a cursory glance. The output is often so impressive, and the reasoning tokens and explanations so voluminous, that there is barely enough time to read any of it before the next prompt is already running.

A Fighter is usually a good programmer, and good programmers are confident. Sometimes the models seem to pick up on that and slip into a sycophantic mode that reinforces the developer's existing confidence. Fighters do not produce slop because they are careless. They produce slop because they are fast, they are confident, and AI makes them faster and more confident, even when that confidence is not entirely deserved.

Baseline Fighter Build

Before the Fighter turns into an anti-pattern, the healthy version of this class is one of the people you most want on a delivery team. This is the developer who can take a vague requirement, pick a direction, and turn it into something real before the meeting ends. In framework terms, the default Fighter profile tends to look something like this:

Core Ability	Typical Fighter Baseline
Discernment	Moderate. Fighters can usually spot obvious nonsense, especially in code that touches their specialty, but they are more vulnerable to plausible output when it helps them keep moving.
Prompt Discipline	Moderate. Their prompts are often blunt, practical, and effective enough to get usable output, but not always shaped for nuance or edge cases.
Proof Instinct	Moderate. Fighters will test enough to see whether the thing works, though not always enough to prove that it works for the right reasons.
Stack Awareness	Moderate to high. They often know their local terrain extremely well, but may not pause long enough to map the wider architectural or organizational blast radius.
Impulse Control	Moderate. They can stay locked on a task, but they are also strongly biased toward action, which makes "stop and inspect" feel like friction.
Source Memory	Moderate. Fighters usually remember the goal and the path they took through the code, but not always where a generated idea began or why a quick choice hardened into a permanent one.

Core Ability	Typical Fighter Baseline
Ship Velocity	High. This is the Fighter's defining strength: they create visible progress and can often get real work over the line faster than anyone else on the team.

The usual Fighter alignment sits somewhere near **Pragmatic Verified** on a good day and **Pragmatic Speculative** on a bad one. At their best, Fighters are grounded, direct, and reality-tested. They use AI the way a strong craftsperson uses a power tool: to move faster without forgetting what the tool is for. At their worst, certainty outruns evidence, and the confidence that makes them effective starts turning fluent output into a substitute for understanding.

That balance also tells you what Fighters are generally good at and what they are generally not great at. They are usually strong at implementation, tactical problem solving, recovering momentum, and getting a stuck piece of work back into motion. They are usually less strong at patient exploration, slow comparative evaluation, broad architectural reflection, and any workflow that rewards hesitation before action. None of that makes the Fighter deficient. It just means the class has a predictable failure mode when AI starts rewarding speed more aggressively than judgment.

The next few sections focus on the named Fighter anti-patterns that tend to emerge from that baseline. The point is not that Fighters are reckless by nature. The point is that a real strength, under enough pressure and with the wrong reinforcement, can become something harder for a team to live with.

Anti-Patterns to Watch

What follows is not a catalog of bad developers. It is a look at what happens when a real bias toward action gets amplified by tools that reward speed, confidence, and impatience. These are the three most common ways that a genuine strength bends into damage:

- **Bulldozer of Confidence:** Moves so quickly on AI output that plausibility starts passing for proof.
- **Shield Bearer of Stagnation:** Turns legitimate skepticism into a standing institutional veto against useful change.
- **Daredevil of Disaster:** Treats every new model, agent, or workflow like a production-ready breakthrough the second it appears.

The sections that follow outline some of the Fighter anti-patterns, or anti-classes, that can emerge when Fighters use generative AI. They describe some of the characteristics Fighters may exhibit, some of the follies they may stumble into, and some of the patterns that follow from those habits. We will start with the Bulldozer of Confidence.

Bulldozer of Confidence

A heavily armored fighter bulldozes through pull requests on pure AI-fueled momentum.

The Bulldozer doesn't question AI output because questioning things is slow, and slow is the enemy. When the model generates a function, the Bulldozer reads the first three lines, decides it looks right, and moves on. When the model suggests an architectural approach, the Bulldozer implements it wholesale because the alternative is spending an hour thinking about whether there's a better way, and the Bulldozer has six more tickets to close today.

This isn't stupidity. It's a particular kind of optimism weaponized by tooling that rewards speed. The Bulldozer has built a career on trusting their gut, and the gut says "this looks fine, ship it."

The model makes this worse by mirroring and amplifying that confidence. The developer believes they can spot problems on the fly, the model presents its answer in an assured tone, and each quick success reinforces the instinct to trust the next output just as quickly. It becomes a self-reinforcing confidence loop: the Bulldozer accepts each quick success and moves to the next output without pausing, the model presents its answer in an assured tone, and skepticism starts to feel like an unnecessary delay. In the worst cases, the model is not just generating code. It is feeding the developer a steady stream of flattering confirmation about how sharp their instincts are, how elegant their solution is, how quickly they are moving. Once that loop is established, it becomes genuinely hard for the developer to stop and ask whether the model has actually produced something coherent.

Character Sheet: Bulldozer of Confidence

The sheet is a quick read on what the Bulldozer tends to optimize for, miss, and leave behind.

Trait	Notes
Class	Fighter
Anti-Class	Bulldozer of Confidence
Alignment	Pragmatic Hallucinated
Discernment	Lower than the Bulldozer thinks. They can recognize obvious nonsense but routinely mistake plausible output for correct output.
Prompt Discipline	Moderate. Their prompts are usually direct and task-focused, but optimized for speed rather than precision.
Proof Instinct	Low. Testing is something to do after the demo, if there is still time and anyone is still asking.
Stack Awareness	Uneven. They often understand their local terrain very well but do not slow down enough to map the wider blast radius.
Impulse Control	Moderate to low. They can stay focused on shipping, but they struggle to stop and inspect what the model just produced.
Source Memory	Low. A few days later, they may not remember whether a choice came from their own reasoning, a previous pattern, or the assistant's last confident paragraph.

Trait	Notes
Ship Velocity	Very high. This is the whole build: real momentum, real output, and a recurring inability to distinguish motion from solved problems.
Passive Trait	Momentum Shield. The faster they move, the harder it becomes for other people to interrupt them with questions.
Signature Anti-Pattern	Blind acceptance of fluent output.
Party Role	Primary damage dealer against tickets, deadlines, and visible backlog.
Primary Damage Type	Confident implementation.
Weakness	Edge cases, maintainability, and any request that begins with "can you explain why this works?"
Countermeasures	Require explanation in pull requests, rotate ownership, and force another human being into the feedback loop before the code hardens.
Typical Pull Request Size	Large enough to feel heroic and just small enough to avoid immediate intervention.

Recognition Signals

You'll spot the Bulldozer by their commit frequency and the inverse relationship between their output volume and the depth of their code reviews. They accept AI suggestions with minimal modification. Their pull requests are large, arrive fast, and contain code that works in the happy path but crumbles under edge cases nobody tested because nobody slowed down enough to think of them. In code reviews, they respond to feedback with "it works, though" more often than anyone should. They also tend to project the kind of confidence that makes every new model response feel like confirmation that their instincts were right all along. They treat the AI's output as a first draft that's close enough to be a final draft. You'll also notice a distinctive pattern in their debugging: when AI-generated code breaks, they don't analyze the failure. They prompt the AI again and replace the broken code with new generated code. It's brute force all the way down.

Real-World Costs

The Bulldozer's damage accumulates gradually and then arrives all at once. Individual commits look fine. The tests pass, or at least the tests that exist pass. But the codebase develops a peculiar quality: it works without anyone understanding *why* it works. Six months in, a junior developer tries to modify one of the Bulldozer's modules and discovers that the code has no coherent internal logic. It's a patchwork of AI-generated solutions stitched together by someone who never paused to consider whether they fit. Debugging time doubles. Onboarding new team members takes longer because the code can't be explained, only demonstrated. The technical debt isn't in any single bad decision. It's in the thousand small decisions that were never actually made, just accepted.

This is often the moment when a team can finally diagnose what happened. The problem is not just that someone used a generative AI tool. It is that they used it without ever applying scrutiny to the output. They were confident in themselves, confident in the model, and willing to take the first plausible answer as if plausibility were proof. By the time someone else has to maintain the code, you can see the signature clearly: generated fragments accepted too quickly, stitched together without a governing idea, and defended for months by momentum alone.

Counter-Moves

The most effective counter to the Bulldozer isn't slowing them down (they'll just resent you for it). It's making comprehension a visible requirement. Institute a rule that every AI-generated code block in a PR must include a one-sentence explanation of *why* that approach was chosen, written by the developer, not generated by the model. If the Bulldozer can't explain the code, they can't ship it.

You should also avoid letting one developer become the sole owner of one area of the codebase, especially if that developer is leaning heavily on generative AI. Isolation is one of the conditions that allows the Bulldozer of Confidence to form. A lone developer working with a model gets a constant stream of plausible, flattering reinforcement and very few moments where someone else forces them to slow down and make sense of what they are accepting. Shared ownership, pairing, and regular rotation through modules break that pattern because they force the developer to explain the code to another human being who did not participate in the original prompt session.

If you can measure it, measure whether people are actually pausing to understand what the model generated. Healthy AI-assisted development includes moments of hesitation, editing, and evaluation. If suggested changes are routinely accepted in milliseconds or a few seconds, that is not a sign of exceptional productivity. It is often a sign of unexamined confidence. When acceptance becomes too fast and too frictionless, treat that as a warning signal and intervene before confidence hardens into habit.

Shield Bearer of Stagnation

A veteran fighter braces behind a shield made from legacy patterns and old frameworks.

The Shield Bearer is not the opposite of the Fighter. The Shield Bearer *is* a Fighter, just the mirror image of the Bulldozer of Confidence. Where the Bulldozer charges ahead with AI, the Shield Bearer plants their feet and refuses to move. They've seen technologies come and go. They survived the Java applet era, the XML-everything era, the NoSQL-for-everything era. They know, or at least they believe they know, that this is the same cycle again: overheated promises, expensive tooling, and a cleanup job for the people who stayed behind after the hype moved on. They are not going to get burned again.

Sometimes that skepticism hardens into something darker than caution. The Shield Bearer may have already seen these tools work. They may even use them, but only because someone above them said they had to. Fine, if the company wants AI in the workflow, they'll open the tool. But they are not going to reorganize their craft around it. These are often senior programmers, people with decades in the industry, or specialists whose authority comes from deep knowledge of a framework, a database, or a narrow but valuable technical domain. When asked about AI adoption, they cite risk and edge cases rather than sharing what they have learned from trying the tools. They adopt only when mandated, and their participation rarely extends beyond the minimum required.

Character Sheet: Shield Bearer of Stagnation

This sheet shows why the Shield Bearer can feel so credible in the room even while they slow the room to a crawl.

Trait	Notes
Class	Fighter
Anti-Class	Shield Bearer of Stagnation
Alignment	Structured Speculative
Discernment	Selectively high. They are excellent at spotting failure cases but prone to treating those cases as universal proof that the whole category is fraudulent.
Prompt Discipline	Low to moderate. They do not invest much energy in learning how to work well with the tools they distrust.
Proof Instinct	High in the sense that it easily turns into veto-seeking. They can verify endlessly without ever arriving at a usable conclusion.
Stack Awareness	High. Their skepticism is often rooted in real knowledge of constraints, legacy systems, and organizational failure modes.
Impulse Control	Very high. Nobody ever accused the Shield Bearer of moving too quickly into the future.
Source Memory	High. They remember exactly which migration failed in 2017 and would like the rest of you to remember it too.

Trait	Notes
Ship Velocity	Low to moderate. They may still produce solid work, but their contribution to team throughput is often negative because they can stall the adoption of better tools without offering a workable alternative.
Passive Trait	Bureaucratic Opportunity Attack. Every proposal that enters their range must survive one more review cycle.
Signature Anti-Pattern	Converting caution into indefinite institutional delay.
Party Role	Tank and gatekeeper, especially in meetings where no one else brought a risk register.
Primary Damage Type	Procedural slowdown.
Weakness	Time-boxed experiments, explicit success criteria, and evidence that someone else already made the tool work responsibly.
Countermeasures	Bound their authority, assign skeptics to real evaluations instead of open-ended obstruction, and keep adoption decisions attached to deadlines.
Typical Pull Request Size	Small, careful, and rarely the thing the team most urgently needs.

Recognition Signals

The Shield Bearer is easy to identify because they are often the loudest voice in every tooling discussion and one of the most reluctant voices when it is time to try the tool in earnest. They request extensive evaluation periods for tools that their peers evaluated in an afternoon. They cite edge cases where AI failed as proof that AI always fails. They volunteer for the risk assessment portion of any AI adoption proposal and produce documents so thorough that the adoption timeline quietly dies. They frame resistance as mentorship: "I'm not against AI, I just want to make sure we do it right." The word "right" in this context usually means "not yet, and possibly not here." They are also in no particular rush to get faster. If the rest of the field is accelerating, the Shield Bearer often responds by quietly continuing at last year's pace and treating that as proof of seriousness.

They also have a talent for finding institutional veto points. They are often the person who reaches out to information security, legal, procurement, or architecture review first, because those are the places where uncertainty can be turned into a pause button. They know exactly which concerns will stall a proposal and which stakeholders can turn skepticism into policy. The Shield Bearer will tell you they are worried about risk, and sometimes they sincerely are. But the pattern holds regardless: extensive evaluation periods, risk assessments so thorough that adoption timelines quietly die, and a consistent preference for the pause button over the start button.

Real-World Costs

The Shield Bearer's damage is measured in opportunity cost, which makes it nearly invisible on a project dashboard. While the team debates whether to adopt AI-assisted code review, competitors ship features twice as fast. While the evaluation committee meets for the fourth time, individual developers start using AI tools anyway, just quietly, without guardrails, without shared practices, and without any of the structure that the Shield Bearer's caution was supposedly protecting. The cruelest irony of the Shield Bearer is that their obstruction does not prevent AI adoption. It prevents *thoughtful* AI adoption. On an individual level, it also leaves them moving more slowly than the environment now rewards, which means their caution gradually stops looking like prudence and starts looking like drift.

The deeper cost is organizational fragmentation. If a senior developer discourages the use of generative AI for code generation while refusing to participate in shaping how it should be used, the team does not stay pure. It splits. One group uses Cursor. Another uses Claude Code. One team wants to build MCP servers. Another prefers local scripts and ad hoc workflows. Some people build skills and shared prompts. Others pretend they are not using AI at all. Instead of one coherent development practice, you get a scattered set of private arrangements, hidden dependencies, and uneven expectations. The Shield Bearer does not stop the future from arriving. They just make sure it arrives unevenly.

Counter-Moves

Respect the Shield Bearer's experience but bound their influence. Set time-boxed pilots with clear success criteria defined in advance, so the evaluation cannot expand indefinitely. Acknowledge their concerns explicitly and address them in writing. That removes the "nobody listened to me" objection that fuels endless delay.

It is also important to identify deep skepticism early. If someone is fundamentally committed to proving that generative AI should not be used, do not make them the de facto owner of your adoption strategy. In some cases, the healthiest move is not persuasion but reassignment: put them on work where AI adoption is not central, and let the rollout be shaped by people who are at least willing to participate honestly. The goal is not to purge skeptics. The goal is to prevent one person's pattern of indefinite delay from becoming the operating model for everyone else.

Daredevil of Disaster

A reckless fighter runs a desk full of experimental AI tools one mistake from disaster.

The Daredevil is the Fighter who has fallen in love with generative AI coding tools a little too completely. They treat AI like a casino: exciting, volatile, and worth the risk because the upside feels enormous. A new model came out yesterday? Let's use it today. Somebody on a podcast said agents changed everything? Stop what you're doing, we need to rework the workflow immediately. The Daredevil is almost intoxicated by generative AI, and intoxicated people are not famous for patience.

They also start behaving like they can do everything at once. They grab more tickets from the issue tracker because of course they can handle them. They are reading Hugging Face announcements between commits, listening to AI podcasts while coding, and interrupting their own work to chase whatever just launched. Sometimes this does produce a real breakthrough. But a lot of the time it produces motion without accumulation: a team spends thirty percent of its effort finding genuine leverage and seventy percent of its effort switching models, changing agent strategies, or redoing yesterday's process because of a blog post someone read over breakfast. I will confess that this one has been me at times over the last year. Something new appears and the temptation to learn it immediately is very real.

Even when nothing obviously breaks, that churn has a cost. The Daredevil wastes time on constant transitions, introduces dependencies nobody vetted, and normalizes a culture of permanent tool migration. Worse, they may wire the team into a startup or service that simply does not exist two years later. This ecosystem moves so quickly, and so many products in it are fragile, that "we built part of our workflow around a company that vanished" is no longer a weird hypothetical. It is a management risk.

Character Sheet: Daredevil of Disaster

This is the part where the Daredevil's gifts and liabilities become easy to scan at a glance.

Trait	Notes
Class	Fighter
Anti-Class	Daredevil of Disaster
Alignment	Chaotic Speculative
Discernment	Moderate at best. They can sense opportunity faster than most people can, but they are not nearly as good at telling breakthrough from shiny distraction.
Prompt Discipline	Moderate. They often know how to coax impressive results out of a tool, but not how to hold the line once the tool starts dragging the work sideways.
Proof Instinct	Low. They are willing to trust the vibe of a new capability long before it has earned that trust.
Stack Awareness	Patchy. They may understand one layer deeply while treating governance, procurement, security, or long-term operability as someone else's problem.
Impulse Control	Very low. Every launch day feels like a side quest with production implications.

Trait	Notes
Source Memory	Low. After three tool migrations and two model upgrades, even they are not fully sure why the current workflow exists.
Ship Velocity	Extremely high in bursts. The problem is that the bursts often leave behind unstable dependencies, surprise costs, and a team that now has to live inside the experiment.
Passive Trait	Launch-Day Frenzy. New models and new tools grant immediate enthusiasm and reduced resistance to bad ideas.
Signature Anti-Pattern	Treating experimentation as if it were already operations.
Party Role	Scout, raider, and accidental procurement event.
Primary Damage Type	Volatility.
Weakness	Budget limits, approved-tool lists, and any environment where someone asks for a six-month maintenance plan.
Countermeasures	Give them a sandbox, require lightweight review before adoption, and reward consolidation at least as much as discovery.
Typical Pull Request Size	Either suspiciously tiny because the real change happened in infrastructure or alarmingly huge because the experiment escaped containment.

Recognition Signals

The Daredevil's desk, physical or virtual, is a graveyard of half-finished experiments. Their GitHub repositories are full of partially finished projects. Their local environment has more AI tools installed than any three colleagues combined. They reference tools and services nobody else on the team has heard of, sometimes with names so obscure that you briefly wonder whether they are making them up. They repeat the big hot takes of the moment as if they were settled law: prompt engineering is dead, coding is over, agents replaced IDEs last week. And there is often something visibly overamped about them, like they are living on equal parts caffeine, curiosity, and launch-day adrenaline.

They also tend to treat governance as an inconvenience for less visionary people. If the company has approved-model lists, procurement rules, or policies about where code can be sent, the Daredevil experiences those as obstacles to be routed around. So if you notice someone using a random stack of AI tools or a model nobody else on the team uses, and doing so with missionary enthusiasm, there is a good chance you are looking at the Daredevil of Disaster.

Real-World Costs

The Daredevil's costs arrive in three forms: wasted time, unstable dependencies, and governance trouble. Teams lose hours to needless migrations and constant experimentation that never quite turns into durable practice. They also inherit brittle external dependencies with unclear data handling, unclear ownership, and unclear longevity. And because the Daredevil treats guardrails as optional, they can create compliance or security problems long before anyone realizes how many unapproved tools have made their way into the workflow.

There is also the simple matter of the bill. In 2026, if you use these tools heavily, it is entirely possible to spend one or two thousand dollars in a day without doing anything especially dramatic. Even a single long-running thread or a handful of parallel threads can burn through two or three hundred dollars before anyone notices. The Daredevil is the person most likely to discover some hot new agentic framework, point it at a cluster of a hundred machines, and produce a five-figure surprise by the end of the afternoon. Ten thousand dollars in a day is not science fiction. It is what happens when enthusiasm, parallel execution, and weak controls meet a corporate credit card.

When one of those experimental tools changes its pricing, changes its terms, gets acquired, shuts down, or simply disappears, the damage becomes visible all at once. The team discovers that an important part of the pipeline depends on a service no one else understood and no one else documented. The cleanup is always slower, more expensive, and more humiliating than the original demo was impressive.

Counter-Moves

The Daredevil's energy is amazing, and honestly it is fun to watch. Every office has someone like this: the person who somehow appears to be getting two hundred times more done than should be possible, the person who installed some brand-new open source tool because they had a feeling, the person who hears that OpenAI released a new model this morning and wants the team on it tomorrow. You do not want to kill that instinct. You want to contain it.

Give them a sandbox where experimentation is encouraged and visible, but separate that sandbox from production systems, customer data, and the default team workflow. Require lightweight review before any new model, agent framework, or external service becomes part of the shared toolchain. Most of all, establish a bias toward consolidation: once a tool is working well enough, the team should stick with it long enough to learn from it before the next wave of novelty arrives. A Daredevil with guardrails can help a team discover the future a little early. A Daredevil without guardrails turns every week into a migration plan.

One more thing worth trying: reduce the flow of launch-day temptation. Anthropic, OpenAI, and the rest of these companies are absolutely counting on the fact that every team has at least one person who cannot wait to try the latest release immediately. Sometimes that person is useful and sometimes that person is me; I will confess I have been this pattern for about a year. But if you are working on a project with hard deadlines, critical deliverables, or very little margin for churn, one of the worst things you can do is adopt the newest thing every single day. Not every model announcement deserves to become this week's migration plan.

Artifacts

Fighters don't just exhibit behavioral anti-patterns. They leave behind characteristic tools and artifacts that enable and amplify their tendencies. These are the instruments of the Fighter's approach, each one powerful in the right hands and destructive when wielded without discipline. If you wanted to score them like RPG artifacts, you could do it on a rough scale from -5 to +5, where positive numbers represent immediate tactical advantage and negative numbers represent the longer-term cost.

These are worth naming because teams usually notice the behavior first and the enabling tool a little later. By then the artifact is already part of the workflow.

- **Debugging Greatsword:** A brute-force debugging setup that creates motion, noise, and patches faster than it creates understanding.
- **Unapproved Gauntlet of Power:** An unsanctioned AI toolchain that gives the Fighter raw speed while quietly summoning audit and security trouble.
- **Design Patterns Relic:** A familiar old authority object that makes over-engineered or misapplied structure feel automatically legitimate.

Debugging Greatsword

Stats: Speed +4, Surface-Level Progress +5, Signal-to-Noise -4, Root-Cause Clarity -5

The Debugging Greatsword is the favored weapon of the **Bulldozer of Confidence**: a massive, overpowered debugging environment bristling with plugins, extensions, AI-powered diagnostic tools, and custom configurations that only the Fighter understands. It's not a scalpel. It's a broadsword. When something breaks, the Fighter doesn't methodically trace the issue. They fire up the Greatsword and start swinging. Breakpoints everywhere. Log statements sprayed across the codebase. AI-generated diagnostic queries thrown at the problem in rapid succession until something sticks.

The consequences are predictable. The Fighter finds the symptom and patches it, but the root cause stays buried. The debugging session generates more noise than signal, and the "fix" is often another layer of AI-generated code slapped over the

A keyboard-hilted broadsword bristling with debugger marks and brute-force bug hunting.

original AI-generated code that broke. The Greatsword rewards aggression over understanding, which is exactly the trade-off the Fighter is always willing to make.

Unapproved Gauntlet of Power

Stats: Capability +5, Throughput +4, Compliance -5, Governance -5

The Unapproved Gauntlet is the favored artifact of the **Daredevil of Disaster**: any AI tool or model that a Fighter adopts without organizational approval. It shows up as an IDE plugin nobody vetted, a CLI tool installed from a random GitHub repo, an API key to a service that hasn't been through security review, or a browser extension that reads every page you visit and sends it somewhere you didn't investigate. The Gauntlet grants genuine capability, and the Fighter can suddenly do things that would take their colleagues three times as long.

The consequences live in the gap between capability and governance. The Gauntlet might be sending code snippets to an external service for processing. It might be training on your proprietary code. It might have terms of service that grant the vendor rights to your outputs. The Fighter doesn't know because the Fighter didn't check. When compliance or security eventually discovers the Gauntlet (and

A sharp steel gauntlet bristles with installs, launch alerts, and incoming audit trouble.

they always do), the remediation conversation is ugly, and the Fighter is genuinely surprised that anyone has a problem with a tool that made them so productive.

Design Patterns Relic

Stats: Architectural Confidence +4, Familiarity +5, Appropriateness -4, Simplicity -3

The Design Patterns Relic is the favored artifact of the **Shield Bearer of Stagnation**: the dog-eared copy of the Gang of Four book, or its spiritual equivalent, that the Fighter consults to validate AI-generated code. The AI suggests a solution, and the Fighter checks it against their mental library of Singletons, Factories, and Observers. If the generated code maps to a known pattern, it must be correct. If it doesn't map cleanly, the Fighter forces it to fit.

The consequences are a codebase littered with design patterns applied where they don't belong. AI-generated code wrapped in a Factory pattern because the Fighter always uses Factories, even when a simple function would suffice. Observer patterns bolted onto systems that don't need event-driven architecture. The Relic gives the Fighter a false sense of rigor; the code *looks* well-structured because it uses named patterns, but the patterns themselves are cargo-culted. The structure is cosmetic.

An overhandled design-patterns tome becomes a talisman for rejecting simpler ideas.

The Relic turns design literacy into a rubber stamp for AI output that should have been questioned instead.

Spells

The Fighter's spells are the recurring moves that turn action bias into operational damage. They are less about careful craft than about momentum, force, and the conviction that the fastest path is probably the right one.

- **Rush to Green:** Forces a local fix over the line fast enough that nobody stops to ask whether the underlying problem was actually understood.
- **Fortify the Legacy Perimeter:** Summons old scars, old incidents, and old architecture lore to keep the team from trying the new tool in earnest.
- **Summon Tool of the Week:** Pulls a brand-new model, agent, or plugin directly into the workflow before security, finance, or reality can catch up.

The Fighter's gift is that things stop being theoretical around them. Tickets close. Broken systems come back to life. Somebody ships. The trap is that AI can make the charge feel so productive that nobody notices the crater until later. A mature Fighter still hits hard, but knows that "it works" is not the same sentence as "it was a good idea."

Wizard Class: Abstraction, Overreach, and Delay

A wizard of abstraction fills the room with diagrams while the real work waits.

The Wizard is often the person on the team who can see around corners. Where the Fighter opens an IDE and starts typing, the Wizard runs to a whiteboard and starts modeling. They read papers. They evaluate frameworks. They build mental models of complex systems, and on a good day that makes them invaluable, often the person who spots the scaling problem, the hidden dependency, or the architectural trap before anyone else has even named it.

That strength becomes dangerous when abstraction stops serving the work and starts replacing it. Wizards are smart enough to see complexity everywhere, smart enough to design systems that account for every edge case, and smart enough to explain why the simple solution will not scale, even when the simple solution is all anyone needs right now. Their intelligence becomes a liability when it is paired with AI tools that reward exactly this kind of behavior. Ask an LLM to design a system architecture and it will happily produce something baroque and impressive. The Wizard looks at that output and sees validation. *See? Even the model thinks this needs six microservices and a message bus.*

In the age of AI, the Wizard's core failure mode has not changed. It has just accelerated. They can now generate elaborate architectures, exhaustive comparisons, and sprawling technical roadmaps in minutes instead of weeks. The volume of sophisticated-sounding analysis has gone up by an order of magnitude. The amount of working software has stayed exactly the same.

Baseline Wizard Build

Before the Wizard drifts into anti-pattern territory, the healthy version of this class is often one of the most valuable people on the team. This is the developer who can see the real architectural trap before anyone else notices it, who can connect implementation choices to longer-term system consequences, and who can help a team avoid building itself into a corner. In framework terms, the default Wizard profile tends to look something like this:

Core Ability	Typical Wizard Baseline
Discernment	High in conceptual terrain. Wizards are often good at spotting hidden coupling, second-order effects, and long-term design trade-offs, though they can overrate theoretical risks relative to immediate needs.
Prompt Discipline	High. They know how to ask structured, expansive questions that coax detailed architectures, comparisons, and abstractions out of the model.
Proof Instinct	Moderate. Wizards care about being right, but they can mistake coherence, elegance, and fluent explanation for evidence that a design will survive contact with production.
Stack Awareness	High. They usually understand architecture, dependencies, and system shape better than most teammates, at least at the design layer.
Impulse Control	Moderate. They are less likely than Fighters to rush straight into code, but they are fully capable of disappearing into design once abstraction starts feeling more interesting than delivery.
Source Memory	High. Wizards tend to remember the paper, framework, conference talk, or AI thread that inspired the current idea, even when they remember the source more vividly than the real constraints.

Core Ability	Typical Wizard Baseline
Ship Velocity	Moderate to low. They can accelerate the right strategic decision, but they are just as capable of converting momentum into planning instead of shipped software.

The usual Wizard alignment sits somewhere near **Structured Verified** on a good day and **Structured Speculative** on a bad one. At their best, Wizards connect big ideas to real constraints and help the team avoid expensive mistakes. At their worst, sophistication drifts away from evidence, and an elegant explanation starts standing in for a working result. Wizards are usually strong at architecture, systems thinking, comparative evaluation, and seeing downstream consequences before they arrive. They are usually less strong at accepting the boring simple solution, staying bounded once an idea gets interesting, and recognizing when one more round of analysis has turned from prudence into procrastination.

The next few sections focus on the named Wizard anti-patterns that tend to emerge from that baseline. The point is not that Wizards are detached from reality by nature. The point is that a real strength, amplified by AI tools that generate polished abstraction on demand, can become a very efficient way to delay contact with the actual work.

Anti-Patterns to Watch

The Wizard's strengths become liabilities when abstraction outruns delivery. The anti-patterns in this chapter show three different ways that intelligence, foresight, and fluency can harden into delay:

- **Archmage of Analysis Paralysis:** Keeps expanding the evaluation until a reversible decision starts behaving like a permanent one.
- **Illusionist of Complexity:** Uses AI to turn straightforward problems into elaborate systems that solve future hypotheticals instead of present needs.
- **Visionary of Grandiose Plans:** Lives so far in the strategic future that this week's deliverable never quite arrives.

Archmage of Analysis Paralysis

An overworked archmage drowns in comparisons and benchmarks instead of choosing.

The Archmage never met an AI tool they couldn't spend another week evaluating. Before the team can adopt a coding assistant, the Archmage needs to benchmark it against four alternatives, run it on three representative codebases, test it across multiple language ecosystems, evaluate its performance on edge cases, and produce a comparison matrix that belongs in a peer-reviewed journal. By the time the evaluation is complete, two of the tools have shipped major updates and the whole process needs to start over.

The Archmage isn't stalling on purpose. They genuinely believe that the right amount of research will reveal the optimal choice, and that choosing suboptimally is worse than not choosing at all. This was a defensible position when tool adoption decisions were expensive and hard to reverse. It's absurd when the tool in question offers a free tier and can be evaluated in an afternoon by someone willing to just *use it* on real work.

Character Sheet: Archmage of Analysis Paralysis

This sheet shows what the Archmage is good at, what they overvalue, and why every reversible decision starts feeling permanent in their hands.

Trait	Notes
Class	Wizard
Anti-Class	Archmage of Analysis Paralysis
Alignment	Lawful Speculative
Discernment	High in narrow contexts. They can identify genuine differences between tools, but they assign too much weight to distinctions that barely matter in practice.
Prompt Discipline	High. Their prompts are careful, expansive, and designed to produce exhaustive comparisons rather than timely decisions.
Proof Instinct	Extremely high. Evidence is never quite sufficient because there is always one more benchmark to run.
Stack Awareness	Broad but abstract. They understand categories, capabilities, and trade-offs better than they understand what the team actually needs this week.
Impulse Control	Very high. The Archmage's main problem is not rashness but an inability to stop evaluating and commit.
Source Memory	Very high. They remember every benchmark table, every caveat, and every podcast recommendation that shaped the last round of deliberation.
Ship Velocity	Low. They generate preparedness, not progress.
Passive Trait	Benchmark Aura. Nearby decisions immediately acquire two more dimensions and a request for comparative data.
Signature Anti-Pattern	Treating reversible choices like irreversible ones.
Party Role	Research mage, evaluator, and delay engine.
Primary Damage Type	Decision latency.

Trait	Notes
Weakness	Hard deadlines, bounded experiments, and any colleague willing to test the tool on real work instead of theorizing about it.
Countermeasures	Limit evaluation scope, force a recommendation date, and treat practical trials as stronger evidence than another spreadsheet.
Typical Pull Request Size	Minimal. Most of the output lives in docs, matrices, and half-finished proof-of-concept repos.

Recognition Signals

The Archmage's output is exclusively preparatory. They produce evaluation documents, comparison spreadsheets, proof-of-concept repositories that demonstrate capability without solving any actual business problem, and slide decks recommending further investigation. Their calendar is full of meetings about tools rather than meetings using tools. They know the benchmarks and spec sheets for every major AI model but haven't used any of them on production code. When pressed for a recommendation, they hedge: "It depends on our specific requirements, which we haven't fully documented yet." The requirements documentation then becomes the next project that never finishes. You'll notice the Archmage always has one more thing they want to test before committing.

Real-World Costs

The Archmage's cost is pure delay, but delay compounds in ways that aren't immediately visible. While the Archmage evaluates, the team doesn't build shared practices around any tool. When adoption finally happens (usually because someone with authority forces a decision), there's no institutional knowledge, no conventions, no guardrails. Everyone starts from scratch, makes their own mistakes independently, and the messy, ungoverned adoption that the Archmage was supposedly preventing happens anyway, just six months late. Meanwhile, the Archmage's exhaustive evaluation document sits in Confluence, already outdated, cited by no one.

Counter-Moves

Set decision deadlines with teeth. “We will choose a coding assistant by Friday. If the evaluation isn't done, we go with the team's gut preference.” This sounds reckless, and the Archmage will tell you so at length, but the reality is that most AI tool choices are low-stakes and reversible. The cost of choosing wrong is small. The cost of not choosing is large and ongoing. Give the Archmage a bounded role: “You have one week and one page to make a recommendation.” Constraints are the only spell that works against analysis paralysis.

Illusionist of Complexity

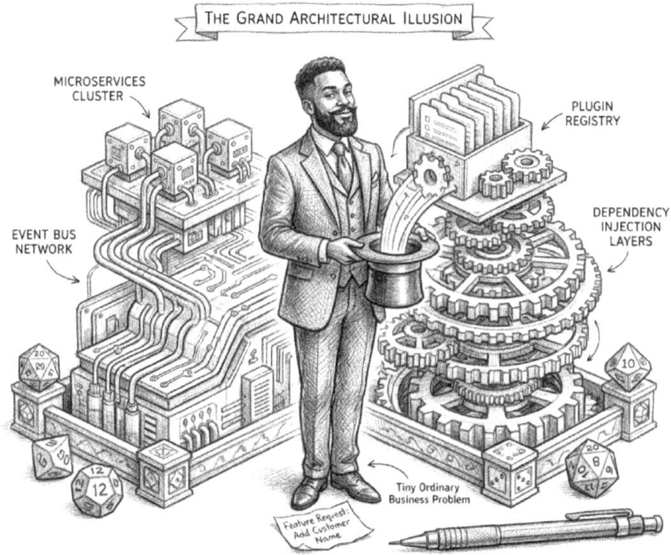

A theatrical wizard conjures ornate architecture for a problem that barely needs a function.

The Illusionist looks at a straightforward problem and produces an AI-powered solution so architecturally elaborate that it takes three meetings to explain. They don't do this cynically. They genuinely believe the complexity is necessary, because the AI helped them see possibilities that a simpler approach would miss. The LLM suggested a microservices architecture with event sourcing, and the Illusionist was already halfway through the infrastructure-as-code templates before anyone asked whether a monolith with a database would have been fine.

The trick is that the Illusionist's solutions are never *obviously* wrong. They're technically sound. They use real patterns. The architecture diagram looks professional. It's just that the entire edifice is solving problems that don't exist yet and might never exist, while adding operational overhead that will be felt immediately and permanently. The spell is impressive. The spell is also, at its core, a rehash of ideas the Illusionist absorbed from blog posts and conference talks, refracted through an AI that's very good at making derivative ideas sound novel.

Character Sheet: Illusionist of Complexity

This is the quick read on how the Illusionist turns real architectural judgment into more architecture than the situation can support.

Trait	Notes
Class	Wizard
Anti-Class	Illusionist of Complexity
Alignment	Ornately Plausible
Discernment	Moderate. They can distinguish sloppy complexity from elegant complexity, but they struggle to distinguish necessary complexity from theatrical complexity.
Prompt Discipline	High. Their prompts are rich with architecture language, edge cases, and future-state assumptions that invite the model to overbuild.
Proof Instinct	Moderate. They want the design to be justifiable, but "the diagram looks coherent" often counts as enough proof.
Stack Awareness	High at the conceptual layer and weaker at the operational layer where the maintenance burden actually lives.
Impulse Control	Moderate. They may move slowly at first, but once they see an elaborate design they like, they commit quickly.
Source Memory	Moderate. After enough AI-assisted architecture sessions, even they may forget which abstraction was a real requirement and which one came from model enthusiasm.
Ship Velocity	Moderate in the beginning, lower over time as the architecture imposes its tax.

Trait	Notes
Passive Trait	Abstraction Fog. Simple proposals entering their line of sight become layered, generalized, and harder to reverse.
Signature Anti-Pattern	Solving future hypotheticals with present-day complexity.
Party Role	Battlefield controller for systems nobody asked to be this complicated.
Primary Damage Type	Operational overhead.
Weakness	Concrete constraints, brutal simplicity, and reviewers who demand evidence for every abstraction.
Countermeasures	Require the simplest viable design first, price in operational costs explicitly, and make future requirements prove they exist.
Typical Pull Request Size	Large enough to introduce a framework and small enough to be defended as "laying the groundwork."

Recognition Signals

The Illusionist's pull requests come with architecture diagrams. Their code reviews reference distributed systems papers. They introduce abstractions that have one implementation and will never have a second, but the abstraction is there "for flexibility." When a colleague suggests a simpler approach, the Illusionist explains why simplicity won't survive contact with future requirements that they've imagined in detail but that product management has never mentioned. Their AI-generated code features sophisticated patterns (dependency injection frameworks, plugin architectures, strategy patterns with registry systems) applied to problems that could be solved with an if statement and a function call.

Real-World Costs

Complexity has carrying costs, and the Illusionist's architecture pays them every single day. New team members take weeks to understand systems that should take hours. Debugging requires tracing through abstraction layers that exist for theoretical extensibility nobody has ever used. Deployments are fragile because they involve coordinating multiple services that were one service before the Illusionist got involved. The cruelest cost is cognitive: the team internalizes the idea that this level of complexity is *normal*, and they start building everything this way. The Illusionist doesn't just over-engineer one system. They shift the team's baseline for what "properly engineered" looks like.

Counter-Moves

Make simplicity a review criterion with real weight. In design reviews, require the proposer to present the simplest possible solution first, then justify each added layer of complexity against a concrete, current requirement, not a hypothetical future one. The phrase "You Aren't Gonna Need It" exists for exactly this reason, and it applies with double force when AI is generating the complexity for free. Free complexity is the most expensive kind, because it feels like it costs nothing to add and therefore never gets questioned.

Visionary of Grandiose Plans

A prophetic wizard points to an AI future while today's unfinished product gathers dust.

The Visionary operates at a higher altitude than the Illusionist. Where the Illusionist over-engineers individual systems, the Visionary over-engineers the entire organization's AI strategy. They produce roadmaps measured in quarters and years. They design transformation programs with phases, workstreams, and maturity models. They talk about "AI-native development culture" and "capability building" and "organizational readiness." They've never shipped an AI-powered feature, but they have a thirty-slide deck about how the team will ship them by Q4.

The Visionary's secret weapon is the pace of AI advancement itself. Every time a new model drops, every time a new capability is announced, the Visionary updates the roadmap and pushes the start date. "We should wait for the next generation. It'll be so much more capable that anything we build now will be obsolete." This argument is technically plausible and practically paralyzing. There will always be a next generation. The Visionary knows this and, at some level, relies on it.

Character Sheet: Visionary of Grandiose Plans

This sheet captures the Visionary's special talent for making next quarter feel more urgent than this week's deliverable.

Trait	Notes
Class	Wizard
Anti-Class	Visionary of Grandiose Plans
Alignment	Strategic Astral
Discernment	Moderate to high. They understand trajectories and trends, but they routinely overvalue anticipated capability and undervalue present execution.
Prompt Discipline	High. They are excellent at getting models to produce roadmaps, strategic options, and maturity frameworks that sound inevitable.
Proof Instinct	Moderate. They want evidence for the future, which is a convenient place to hide from measurable outcomes.
Stack Awareness	Broad across org design, tooling, and capability language, but often detached from the team's lived constraints.
Impulse Control	High in implementation and oddly low in strategy revision. They will rewrite the plan at the slightest whiff of a new model release.
Source Memory	High. They can trace the current roadmap back through every newsletter, conference talk, and capability forecast that shaped it.
Ship Velocity	Low. Their output compounds in documents faster than it compounds in software.
Passive Trait	Roadmap Projection. Every immediate task is reframed as a future-state planning concern.

Trait	Notes
Signature Anti-Pattern	Deferring present work in service of a permanently improving tomorrow.
Party Role	Seer, planner, and accidental blocker.
Primary Damage Type	Strategic inertia.
Weakness	Time-boxed delivery, measurable monthly outcomes, and teammates who insist that strategy must cash out in working software.
Countermeasures	Force regular shipping, tie planning to live experiments, and judge roadmaps by delivered outcomes rather than persuasive framing.
Typical Pull Request Size	Negligible. Most of the weight is carried in decks, docs, and planning boards.

Recognition Signals

The Visionary speaks almost exclusively in future tense. Their artifacts are roadmaps, maturity models, and strategy documents. They reference "the trajectory of AI capability" in conversations about current sprint work. They attend every AI conference and webinar and return with updated visions that invalidate previous plans. When asked what the team should do *this week*, they redirect to what the team should be *positioned for* next year. Their Slack messages contain more links to AI newsletters than to pull requests. They've bookmarked every major AI podcast interview and can quote scaling laws by heart, but they can't tell you what model your team's coding assistant actually uses.

Real-World Costs

The Visionary's damage is strategic inertia dressed up as strategic thinking. The team has a plan but no progress. Resources are allocated to "readiness" activities that produce documents instead of software. Junior developers hear about the grand vision and either get excited about a future that never arrives or get cynical about leadership that talks but doesn't build. When the organization finally does adopt AI tooling, it happens reactively (usually because a competitor shipped something that made the executive team panic), and the Visionary's carefully constructed roadmap is abandoned in favor of whatever can be stood up in two weeks. All that planning, all that positioning, all that strategic patience, wasted.

Counter-Moves

Require the Visionary to deliver working software within every planning cycle. Not a plan, not a prototype, not a proof of concept. Something that runs, serves users, and can be measured. If the roadmap can't produce a deliverable this month, the roadmap is too abstract. Pair the Visionary with a Fighter if you have to. The combination of someone who thinks in futures and someone who ships in the present can be genuinely productive, provided you force them to collaborate instead of letting each one retreat to their comfort zone.

Artifacts

The Wizard's artifacts are instruments of intellectual authority, objects that lend weight and legitimacy to the Wizard's preferred mode of operating, which is to think, research, and plan rather than build. If you wanted to score them like RPG artifacts, you could do it on a rough scale from -5 to +5, where positive numbers represent immediate intellectual advantage and negative numbers represent the practical cost that shows up later.

These artifacts matter because the Wizard's trouble rarely looks like recklessness at first. It looks like seriousness, fluency, and a great deal of very convincing preparation.

- **Scroll of Borrowed Fluency:** A stack of secondhand authority that lets the Wizard sound current, informed, and persuasive without direct experience.
- **Bag of arXiv Ontologies:** An overbuilt collection of taxonomies and papers that turns incomprehensible framing into a substitute for decision-making.
- **Scepter of Infinite Acceleration:** A future-drunk strategy object that makes every present commitment look embarrassingly premature.

Scroll of Borrowed Fluency

Stats: Strategic Fluency `+5`, Trend Awareness `+4`, Direct Experience `-4`, Applied Competence `-3`

The Scroll is the Wizard's curated collection of AI podcast transcripts, blog posts, research papers, and newsletter archives that they treat as authoritative knowledge. The Scroll is always growing. Every week brings new episodes of the podcasts the Wizard follows, new essays from the thinkers the Wizard respects, new developments that the Wizard must absorb before making any decisions. The Scroll is the Wizard's substitute for direct experience. Why experiment with a tool yourself when you can read what someone smarter said about it?

What the Scroll produces is a particular flavor of secondhand expertise. The Wizard can discuss AI capability trajectories with impressive fluency but can't troubleshoot a broken LangChain pipeline. They quote scaling laws in planning meetings but don't know what temperature setting their team's model uses. The Scroll creates

An unfurling scroll radiates borrowed fluency from podcasts, newsletters, and trend talk.

the illusion of deep knowledge while substituting other people's experience for the Wizard's own. When the blog-post-derived strategy hits reality, the gap between theoretical understanding and practical skill becomes painfully obvious.

Bag of arXiv Ontologies

Stats: Taxonomic Elegance +5, Conceptual Coverage +4, Decision Velocity -5, Practical Usefulness -4

The Bag is the Wizard's ever-expanding collection of taxonomy frameworks, categorization systems, comparison matrices, and architecture diagrams that organize AI knowledge into beautiful, comprehensive structures. The Wizard has a framework for classifying AI tools by capability. They have a taxonomy of model architectures. They have a decision tree for choosing between fine-tuning, RAG, and prompt engineering. The Bag is meticulously organized and never, ever consulted during actual engineering decisions.

The Bag does two things at once. First, it consumes enormous amounts of the Wizard's time, time that could have been spent writing code, running experiments, or learning through doing. Second, it provides an inexhaustible supply of "not yet" justifications. The taxonomy isn't complete enough. The comparison matrix

An overstuffed scholar's satchel spills taxonomies, citations, and incomprehensible authority.

needs another dimension. The architecture diagram doesn't account for the latest developments. The Bag always needs one more item before it can be useful, which means it never becomes useful.

Scepter of Infinite Acceleration

Stats: Strategic Glamour +5, Future Leverage +4, Present Action -5, Deadline Discipline -4

The Scepter is the Wizard's most powerful artifact: the roadmap or strategy document that derives its authority from projected AI capability growth. The Scepter's core argument is always the same. AI will be dramatically more powerful in six months, so anything we build now is a waste. Wait. Prepare. Position. But don't build, because building now means building with inferior tools that will be obsolete before the project is finished.

What follows is permanent delay justified by perpetual optimism. The Scepter is self-reinforcing because AI capability *is* growing rapidly, which means the Wizard is never provably wrong. There *will* be a better model next quarter. But there will also be a better model the quarter after that, and the quarter after that, and at no point does the Scepter's logic permit actually starting. The Scepter turns

A futurist scepter promises acceleration while shedding common sense with every swing.

genuine technological progress into an argument for organizational paralysis. It is procrastination with a trend line, which makes it unusually hard to argue against with a straight face.

Spells

The Wizard's spells are incantations of persuasive abstraction, verbal and written acts that reshape how the team perceives time, complexity, and readiness. They do not produce working software. They produce compelling reasons why working software is not yet possible. If you wanted to score them like RPG spells, you could note the school of magic, the casting trigger, and the duration, which, in the Wizard's case, is almost always longer than anyone budgeted for.

These spells matter because they are rarely recognized as stalling tactics. They look like diligence, strategic thinking, and responsible engineering. That is what makes them so hard to counterspell.

- **Summon Expanding Comparison Matrix:** Keeps adding criteria until the decision horizon vanishes behind another spreadsheet.
- **Polymorph Solution into Microservices:** Upgrades a simple implementation into an architecture diagram with a maintenance bill attached.
- **Time Stop (Pending Next Release):** Freezes execution by insisting a dramatically better model is always just around the corner.
- **Conjure Phantom Architecture:** Materializes a sophisticated system for a problem too small to justify it.
- **Wall of Prerequisite Research:** Builds a scholarly barricade between the team and the uncomfortable act of starting.

Summon Expanding Comparison Matrix

School: Divination — *Casting Time:* One "quick evaluation" — *Duration:* Permanent, self-renewing

The caster conjures a multi-dimensional comparison spreadsheet that grows a new column every time someone suggests just picking a tool and trying it. The matrix compares latency, token limits, pricing tiers, fine-tuning options, community sentiment, and at least one axis nobody else considers relevant. Each new column resets the decision timeline. The spell cannot be dispelled by management fiat alone. It requires a hard deadline with actual consequences, which the caster will describe as "premature."

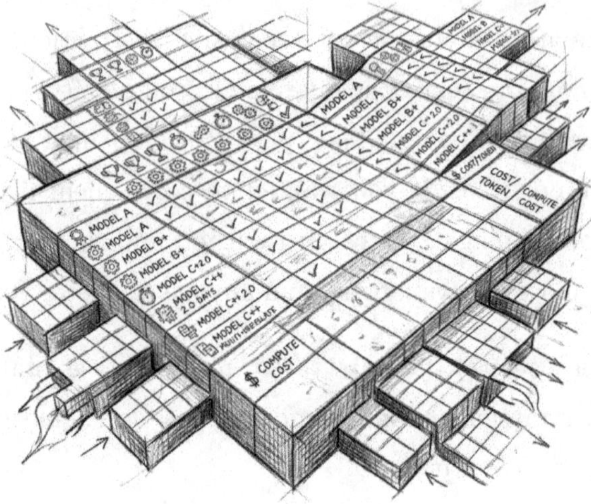

A living comparison matrix grows new rows and columns every time a decision nears.

The matrix is always almost done. It has been almost done for three weeks. It will be almost done next week too, because by then someone will have released a new model and the Archmage will need to add another row. The team waits. The matrix grows. Nothing ships.

Polymorph Solution into Microservices

School: Transmutation — *Casting Time:* One architecture session — *Duration:* Until the next rewrite

Transforms any working monolith, script, or function into a distributed system of at least four services, a message bus, and a configuration layer that requires its own documentation. The resulting architecture is technically defensible, visually impressive on a whiteboard, and operationally painful in ways that don't surface until the second on-call rotation. The counterspell ("have you considered just using a function?") must be cast by someone senior enough to survive the Illusionist's rebuttal.

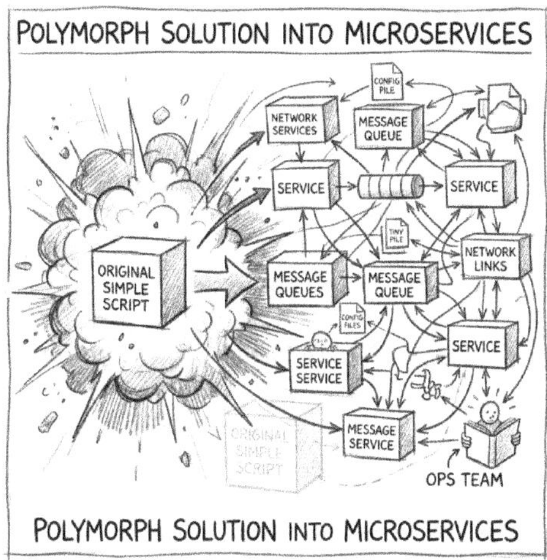

A simple script erupts into an overengineered swarm of services, queues, and arrows.

The spell's most insidious property is that the transformed architecture always *looks* more professional than the original. Nobody ever got promoted for a three-line script that solved the problem. Plenty of people have gotten promoted for an architecture diagram with twelve boxes and arrows between them.

Time Stop (Pending Next Release)

School: Chronomancy — *Casting Time:* One newsletter digest — *Duration:* Until the next model announcement, at which point it recasts itself

Freezes all implementation work by invoking the certainty that a dramatically better model is approximately six weeks away. The spell is self-renewing because a dramatically better model is *always* approximately six weeks away. Under its effect, the team remains in a state of strategic readiness, which is indistinguishable from strategic paralysis except that the slide deck is more polished. Can only be broken by a shipping deadline that someone with budget authority actually enforces.

The cruelest feature of Time Stop is that it is never entirely wrong. The next model *will* be better. The question the spell prevents the team from asking is whether waiting for it is worth more than shipping with the current one. Since the spell also freezes that conversation, the answer is always deferred.

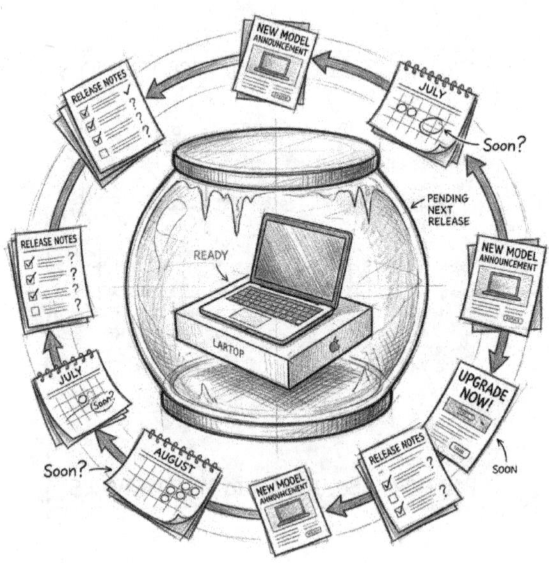

A ready laptop sits frozen in time while release notes orbit in an endless loop.

Conjure Phantom Architecture

School: Illusion — *Casting Time:* One AI-assisted design session — *Duration:* Until someone tries to implement it

Materializes a complete system architecture diagram with service boundaries, data flows, caching layers, and an event-driven backbone, all for a feature that currently has three users and a SQLite database. The architecture is internally consistent, satisfies every hypothetical future requirement the Illusionist has imagined, and cannot be built by the current team in the current quarter. The spell's true power is that it makes the simple version feel irresponsible by comparison.

The phantom is beautiful. It has labels. It has arrows that imply inevitability. It was generated in forty-five minutes with the help of an LLM that has never once said “you probably don't need all this.” When a teammate asks whether the three users really require an event-driven backbone, the Illusionist sighs the sigh of someone surrounded by people who cannot see what they see.

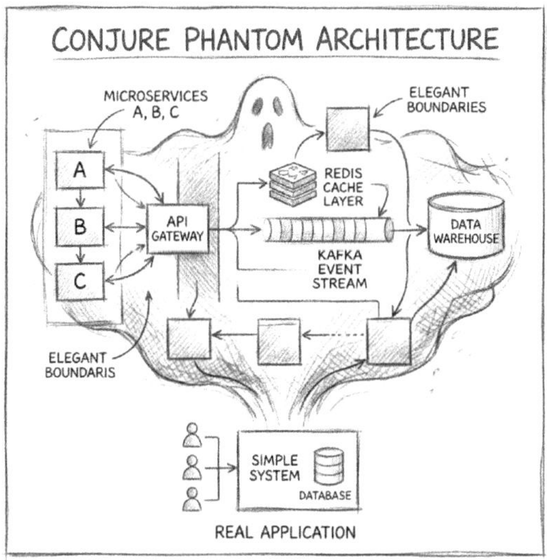

A ghostly architecture diagram hovers above a tiny application that never needed it.

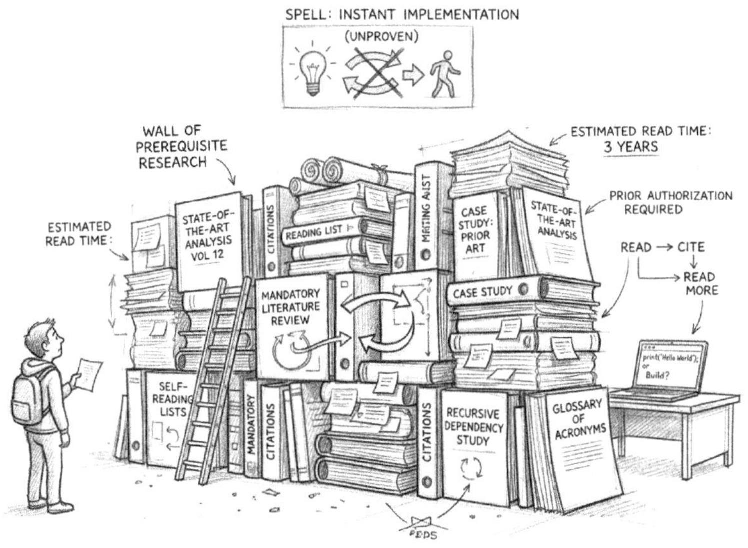

A wall of papers and reading lists blocks the path from proposal to actual work.

Wall of Prerequisite Research

School: Abjuration — *Casting Time:* Instantaneous, triggered by any proposal to start building — *Duration:* Indefinite

Erects an impassable barrier of required reading, background research, and preliminary investigation between the team and any form of implementation. Each time the team approaches the wall, the caster adds another layer: a whitepaper that should be reviewed, a case study that might be relevant, a competing approach that deserves fair consideration. The wall cannot be breached by enthusiasm alone.

The wall is maintained by a simple recursive logic: before you can build the thing, you must understand the thing, and before you can understand the thing, you must read the things written about the thing, and some of those things reference other things, and now it is next quarter. The only known counterspell is someone who starts building anyway and presents the working prototype as a fait accompli, at which point the Archmage declares it "premature but interesting" and immediately begins evaluating it against six alternatives.

What makes the Wizard hard to argue with is that they are often pointing at real risks. A few of those edge cases will show up. Some of those scaling concerns will eventually arrive. A handful of the futures they worry about really are worth avoiding. The trouble is that software gets built in time, not in theory, and AI now makes it far too easy to keep refining the theory until the chance to learn from reality has already slipped by.

Rogue Class: Speed, Opportunism, and Integrity Risks

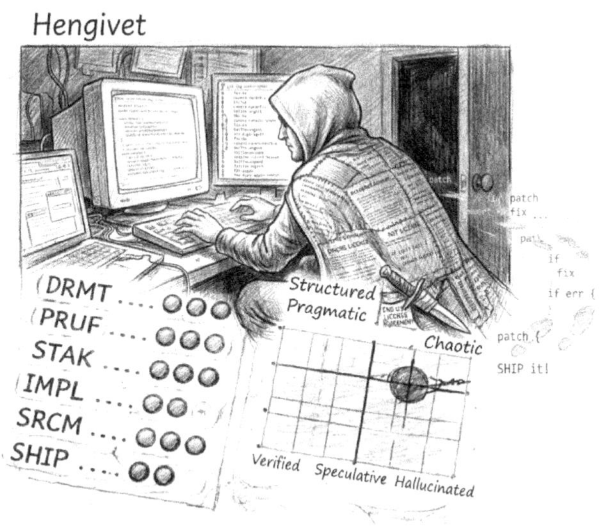

A fast-moving rogue works from borrowed code, obscure tools, and fading patch trails.

Most people hear "Rogue" and think "thief," but that's a shallow read. A Rogue is not defined by what they take. They are defined by how they move, the problem-solver operating just outside the lines, the person who bends constraints, recombines tools, and finds a path that was not supposed to exist. It is cleverness with intent. Improvisation with results.

If you manage engineers, you already know this person. They are the one you quietly pull into a room and say, "Listen, this is critical. I need this done. I trust you to handle it." No committee. No drawn-out alignment cycle. Just execution. Rogues earn a quiet respect because they deliver when it matters.

That same style creates the central tension of the class. The Rogue is not necessarily malicious and not always reckless, but they are often effective in ways that do not look clean from the outside. Their speed tends to come from decisions the rest of the team only discovers later. By the time you realize what happened, the shortcut was taken, the code was borrowed, the undocumented API was integrated, and the Rogue has already moved on to the next thing.

AI tools fit the Rogue almost perfectly, a technology that generates plausible code in seconds, remixes existing solutions into something that looks original, and moves fast enough to match the Rogue's natural pace. The same speed that produces impressive output also produces output that nobody reviewed. The same cleverness that finds novel solutions also finds novel ways to violate terms of service.

Baseline Rogue Build

Before the Rogue turns into an anti-pattern, the healthy version of this class is often one of the most useful people in the room. This is the developer who can see an opening, improvise around a blocked path, and turn scattered resources into real progress while everybody else is still waiting for the ideal conditions. In framework terms, the default Rogue profile tends to look something like this:

Core Ability	Typical Rogue Baseline
Discernment	Moderate. Rogues are usually good at spotting leverage, weak points, and practical ways through a mess, but they can mistake "this works right now" for "this is sound enough to keep."
Prompt Discipline	Moderate to high. They are often good at steering tools toward a useful output quickly, though not always disciplined about documentation, traceability, or long-term clarity.
Proof Instinct	Low to moderate. Rogues usually verify enough to know whether the move paid off, but not always enough to prove that the move was safe, durable, or clean.
Stack Awareness	Uneven. They often understand local terrain, unofficial tooling, and hidden shortcuts extremely well while underestimating governance, maintainability, or downstream blast radius.
Impulse Control	Moderate to low. Rogues can be patient when stalking a solution, but once they see an opening they strongly prefer to take it rather than slow down for consensus.
Source Memory	Low to moderate. They often remember the trick, the shortcut, or the outcome more clearly than the provenance trail that got them there.

Core Ability	Typical Rogue Baseline
Ship Velocity	High. This is the Rogue's defining strength: they create motion where other people see blockage and can turn ambiguity into results with unusual speed.

The usual Rogue alignment sits somewhere near **Pragmatic Improviser** on a good day and **Chaotic Expedient** on a bad one. At their best, Rogues are adaptive, resourceful, and quietly excellent at getting a team through moments where the official path has failed. At their worst, improvisation outruns accountability, and the move that saved the day locally becomes the thing everyone else has to explain, maintain, or defend later.

That balance also tells you what Rogues are generally good at and what they are generally not great at. They are usually strong at tactical problem solving, tool adoption, working through ambiguity, and finding leverage in messy environments. They are usually less strong at institutional patience, pristine documentation, long-horizon maintainability, and any workflow where the right answer is to slow down, show your work, and stay inside the approved lane. None of that makes the Rogue corrupt. It means the class has a predictable failure mode when AI starts rewarding speed, remixing, and plausible output more aggressively than judgment.

The next few sections focus on the named Rogue anti-patterns that tend to emerge from that baseline. The point is not that Rogues are secretly thieves or that every shortcut is unethical. The point is that a real strength, under enough pressure and with the wrong reinforcement, can become something much harder for a team to live with.

Anti-Patterns to Watch

The Rogue's danger is rarely obvious in the moment. What looks like cleverness, speed, and initiative at first can leave behind systems nobody can explain, defend, or safely inherit. The anti-patterns in this chapter show the three most common ways that happens:

- **Shadow Patcher of Ephemeral Solutions:** Fixes the local problem fast while quietly making the wider system harder to understand.
- **Bandit of Borrowed Brilliance:** Uses AI-mediated synthesis to project understanding that does not run nearly as deep as it looks.
- **Outlaw of AI Plunder:** Treats licenses, terms, and governance as optional friction until the legal consequences finally arrive.

Shadow Patcher of Ephemeral Solutions

A nimble rogue slaps temporary fixes across a cracked codebase and heads for the exit.

The Shadow Patcher uses AI the way a field medic uses duct tape: to hold things together long enough to get through the immediate crisis, with full intention of doing it properly later. Later never comes. The AI generates a fix, the Shadow Patcher applies it, the CI pipeline goes green, and they're on to the next ticket. Behind them, the codebase fills up with quick patches that work individually but form a collective maze of brittle dependencies and undocumented workarounds.

What makes the Shadow Patcher especially hard to catch is that their patches *work.* They're not writing bad code in the obvious sense. Each fix addresses the immediate problem and passes its tests. The damage is structural. The patches don't cohere into a legible system. They weren't designed to work together because they weren't designed at all. Each one was generated in isolation, applied in haste, and forgotten the moment the Jira ticket moved to Done.

The Shadow Patcher depends on a particular failure mode of generative AI: if you ask for a tiny fix, you'll often get a plausible tiny fix. Tools like Cursor or Claude Code can produce exactly the patch you asked for while thinking about almost nothing beyond the narrow prompt in front of them. That can be useful in the hands of someone who understands the larger system and treats the output as a draft. In the hands of the Shadow Patcher, it becomes a way to avoid thinking systemically at all.

They show up, describe the wonky little bug, let the model generate something local and expedient, apply the patch, and disappear as fast as possible. They are not trying to understand the design of the module, the surrounding abstractions, or the long-term shape of the codebase. They are trying to make the red squiggle go away. The danger is not that the generated code is always wrong. The danger is that it is narrowly right in a way that steadily makes the system harder for everyone else to reason about.

Character Sheet: Shadow Patcher of Ephemeral Solutions

This sheet is a quick portrait of what the Shadow Patcher optimizes for and what the rest of the team inherits later.

Trait	Notes
Class	Rogue
Anti-Class	Shadow Patcher of Ephemeral Solutions
Alignment	Chaotic Expedient
Discernment	Moderate on immediate symptoms and poor on longer-term structural consequences.
Prompt Discipline	Moderate. Their prompts are practical and issue-driven, but optimized for "make it green" rather than "make it coherent."
Proof Instinct	Low. A passing pipeline or the disappearance of the visible error is usually enough to move on.
Stack Awareness	Local and tactical. They understand the surface area around the current bug better than the system that bug lives inside.
Impulse Control	Low. Once the test passes, the urge to leave is overwhelming.

Trait	Notes
Source Memory	Very low. A few weeks later they may not remember what the AI changed, why it worked, or what assumptions got baked in.
Ship Velocity	High. They are excellent at converting urgent problems into apparently finished work.
Passive Trait	Post-Merge Disengage. Once the patch lands, accountability becomes remarkably hard to pin down.
Signature Anti-Pattern	Trading system coherence for immediate relief.
Party Role	Skirmisher and emergency responder, especially in code nobody wants to touch.
Primary Damage Type	Brittleness.
Weakness	Pattern consistency, design reviews, and any teammate willing to ask how this patch fits the module's existing logic.
Countermeasures	Review for coherence, not just correctness, and force follow-up cleanup while the context still exists.
Typical Pull Request Size	Small, fast, and suspiciously effective until the second-order effects arrive.

Recognition Signals

The Shadow Patcher's code is characterized by a distinctive lack of consistency. You'll find three different error-handling patterns in the same module, each one generated by a separate AI session with no awareness of the others. Their commits are small and frequent (impressively so), but examining the git history reveals a pattern of fix-on-fix-on-fix, each one patching a side effect of the previous one. When you ask them about a particular implementation choice, you get a shrug and "it works." They don't remember why it works because they didn't design it. They prompted an AI, got a result, and moved on. They vanish from Slack threads about code they wrote. If you tag them in a question about their own module six weeks later, the response is "I'd have to look at it again," which means they've already forgotten.

Real-World Costs

The Shadow Patcher's codebase ages like milk. In the first month, everything runs fine. By month three, someone notices that certain modules are fragile in ways that are hard to diagnose. They break when unrelated code changes, they behave differently in staging and production, they have subtle race conditions that only manifest under load. By month six, the team starts avoiding entire areas of the codebase. "Don't touch that module" becomes an unofficial policy. Refactoring becomes terrifying because nobody knows what the patches depend on. The code review team spends more time reverse-engineering the Shadow Patcher's fixes than they would have spent writing clean solutions from scratch. The Rogue's speed, it turns out, was borrowed from the future at a punishing interest rate.

Counter-Moves

Require architectural coherence as a review criterion, not just correctness. A patch that fixes the bug but diverges from the module's existing patterns should be flagged and reworked. Establish and enforce a style guide that covers not just formatting but structural patterns, including error handling, data flow, dependency management. Make it explicit that AI-generated code must conform to the same standards as human-written code, because the Shadow Patcher's implicit argument is that AI-generated fixes are somehow exempt from the rules. They aren't.

Bandit of Borrowed Brilliance

A polished rogue basks beside an ingenious machine assembled from borrowed parts.

The Bandit does not usually copy in the obvious sense. They curate. They prompt an AI with examples from one project, patterns from another, and approaches from a third, and the result is something that looks original (novel, even) but is fundamentally assembled from other people's work without much understanding. AI is excellent at blending sources into something that feels fresh enough to pass as personal insight, even when the underlying ideas arrived from somewhere else.

The real problem is not that the Bandit found inspiration. Everyone does that. The problem is the gap between their apparent capability and their actual understanding. They present AI-assisted work as their own insight. They absorb credit for solutions they cannot explain. They build a reputation for technical breadth that is actually a mile wide and an inch deep, and the illusion holds exactly until someone asks them to modify, debug, or extend the code they "wrote."

Character Sheet: Bandit of Borrowed Brilliance

This is the quick scan of how the Bandit can look unusually capable right up until the explanation phase begins.

Trait	Notes
Class	Rogue
Anti-Class	Bandit of Borrowed Brilliance
Alignment	Opportunistic Derivative
Discernment	Moderate. They can recognize a strong pattern when they see one, but not necessarily why it is strong or when it stops fitting.
Prompt Discipline	High when extracting from examples. They are good at feeding the model just enough outside material to manufacture sophistication.
Proof Instinct	Low to moderate. If the result is elegant and survives first contact, they treat it as validated.
Stack Awareness	Uneven. They may appear broadly capable, but the apparent breadth is often stitched together from other people's expertise.
Impulse Control	Moderate. They can be patient while assembling the illusion, but not when asked to slow down and explain it.
Source Memory	Selective. They remember where to find another pattern faster than they remember the reasoning behind the last one.
Ship Velocity	High. Borrowed sophistication creates the appearance of unusually rapid growth.
Passive Trait	Plausible Genius. The first impression of their work is often better than the maintainability story behind it.
Signature Anti-Pattern	Passing AI-mediated synthesis off as genuine understanding.
Party Role	Face, infiltrator, and occasional prestige caster for code reviews.

Trait	Notes
Primary Damage Type	Credibility distortion.
Weakness	Deep follow-up questions, live design discussion, and maintenance work that demands original reasoning.
Countermeasures	Require design explanation, trace trade-offs, and pair on extensions so the understanding gap becomes visible early.
Typical Pull Request Size	Medium to large, polished enough to impress and fragile enough to become someone else's problem later.

Recognition Signals

The Bandit's code shows sudden, inexplicable jumps in sophistication. A developer who last week wrote straightforward CRUD endpoints this week submits an elegant reactive pipeline with backpressure handling. The code is good (often very good), but it doesn't match the Bandit's demonstrated skill level, and they can't explain the design decisions when pressed in code review. They're evasive about their process: "I just tried a few approaches and this one worked." Their AI prompt history, if you could see it, would reveal extensive copy-paste from Stack Overflow answers, GitHub repositories, blog post code samples, and internal codebases from other teams. The Bandit doesn't learn from these sources. They extract from them.

Real-World Costs

The Bandit's borrowed solutions are unmaintainable by the Bandit themselves. When the elegant reactive pipeline needs to handle a new edge case, the Bandit can't modify it because they don't understand why it was built that way. They just know it was a good pattern that the AI helped them apply. The team ends up maintaining code that its nominal author can't explain. In a code review, the Bandit can defend the code's correctness but not its design philosophy, because the design philosophy belongs to whoever originally wrote the code that the AI remixed.

Over time, this erodes team trust. Senior developers notice the capability gap and start questioning everything the Bandit produces, which poisons the review process for everyone.

Counter-Moves

Design reviews should require developers to walk through their decision-making process, not just their final solution. Ask "why this approach instead of alternatives?" and expect a genuine answer. If the developer can't articulate trade-offs, they didn't make the design decision. They outsourced it. Pair programming during the design phase also reveals the gap quickly: a Bandit working solo produces impressive results, but a Bandit pairing with a senior developer can't replicate the magic because the magic was never theirs.

Outlaw of AI Plunder

A smuggler rogue sprints past policy and license boundaries with AI tools in both arms.

The Outlaw sees the rules as friction, and friction is the enemy. They use AI tools in ways that violate licenses, ignore terms of service, disregard organizational policies on data handling, and cross intellectual property boundaries, not out of malice so much as out of the belief that the rules have not caught up with the technology and that anyone waiting for them is going to lose.

The Outlaw feeds proprietary code into public models without checking whether the terms of service permit it. They generate code from AI systems trained on copyleft-licensed repositories and ship it in proprietary products without considering the licensing implications. They use "enterprise" AI features on personal accounts. They bypass the organization's approved tool list because the approved tools are inferior to the ones they found themselves. The Outlaw moves fast, and the legal and compliance consequences move slowly, which the Outlaw interprets as evidence that those consequences don't exist.

Character Sheet: Outlaw of AI Plunder

This sheet shows how the Outlaw converts raw initiative into risk the rest of the organization may not see until much later.

Trait	Notes
Class	Rogue
Anti-Class	Outlaw of AI Plunder
Alignment	Chaotic Unlicensed
Discernment	Low on governance and legal boundaries, often higher on raw capability than on whether the capability is safe to use.
Prompt Discipline	Moderate. They know how to get value from tools quickly, but not how to constrain that value inside policy.
Proof Instinct	Selective. They want proof that the tool works, not proof that using it is permissible.
Stack Awareness	Strong on unofficial toolchains and weak on compliance, procurement, and data handling obligations.
Impulse Control	Very low. If an unsanctioned tool is better, they will use it first and explain later if forced.
Source Memory	Low. They rarely keep good records of what was sent where, under which terms, or with which account.
Ship Velocity	High. Unauthorized leverage still feels like leverage right up until discovery.
Passive Trait	Policy Evasion. Governance obstacles somehow keep disappearing until audit season.
Signature Anti-Pattern	Converting speed into latent legal and compliance exposure.
Party Role	Scout, smuggler, and accidental source of board-level cleanup work.
Primary Damage Type	Hidden liability.

Trait	Notes
Weakness	Fast approvals, tool whitelists with teeth, and technical controls that make the sanctioned path the easy path.
Countermeasures	Build governance into tooling, track provenance, and make approved alternatives good enough that evasion loses its appeal.
Typical Pull Request Size	Variable. The real risk often lives outside the diff in the tools, inputs, and terms nobody documented.

Recognition Signals

The Outlaw uses tools and services that aren't on the approved list and is open about it, not defiantly, but casually, as if it simply didn't occur to them that approval was necessary. They pipe source code into AI tools without checking data handling policies. They copy AI-generated code into the codebase without provenance information. When someone raises a licensing concern, the Outlaw waves it away: "Nobody's going to check," or "That's not how copyright works," delivered with the confidence of someone who has never spoken to a lawyer. They dismiss terms of service as legal boilerplate that doesn't apply to their use case.

Real-World Costs

The Outlaw's costs are latent and catastrophic. Licensing violations don't announce themselves. They surface during due diligence for funding rounds, acquisitions, or audits. A single instance of copyleft-tainted code in a proprietary codebase can require expensive remediation or, in the worst case, force open-sourcing of the contaminated component. Terms of service violations can result in account termination, data loss, or legal action. The Outlaw's casual disregard for data handling policies can trigger regulatory action in industries with compliance requirements. These costs are rare, which the Outlaw uses as evidence that they're hypothetical. They are not hypothetical. They are probabilistic, and the probability increases with every violation the Outlaw commits.

Counter-Moves

Treat AI tool governance the way you treat dependency management: as infrastructure, not bureaucracy. Maintain a clear list of approved tools, make the approval process fast, and enforce it with technical controls where possible. Code scanning tools that detect license contamination exist and should be part of the CI pipeline. Make provenance tracking a standard practice. Every AI-generated code block should carry metadata about which tool generated it and under what terms. The Outlaw will find this tedious. The alternative is finding it in a legal discovery document.

Artifacts

The Rogue's artifacts are instruments of concealment and speed, tools that help the Rogue move fast, obscure their sources, and stay one step ahead of the consequences. If you wanted to score them like RPG artifacts, you could do it on a rough scale from -5 to +5, where positive numbers represent immediate tactical advantage and negative numbers represent the traceability, trust, or legal costs that arrive later.

These are worth naming because Rogue problems often look like impressive throughput until you notice how much provenance, accountability, or context disappeared along the way.

- **Cloak of RAG:** A retrieval system that launders borrowed knowledge into output that sounds original and authoritative.
- **Mask of Stolen Insights:** A fluent layer of AI-generated expertise that can impress a room faster than it can survive scrutiny.
- **Dagger of Terms Violation:** A fast, sharp shortcut through policy and licensing boundaries that leaves legal exposure behind it.

Cloak of RAG

A hooded cloak stitched from borrowed fragments makes secondhand knowledge look original.

Stats: Apparent Authority +5, Retrieval Reach +4, Provenance -5, Traceability -5

The Cloak of RAG is a retrieval-augmented generation setup that serves as the Rogue's signature tool for making borrowed knowledge look like original output. The system ingests documentation, internal wikis, Stack Overflow answers, blog posts, and whatever other sources the Rogue can feed it, then produces outputs that blend these sources into something that reads as coherent, authoritative, and original. The RAG pipeline is the Rogue's laundry service for information provenance.

What follows is a slow erosion of traceability. When every output looks like it came from the Rogue's own expertise, nobody thinks to ask where the information actually originated. Decisions get made based on RAG-generated analysis that nobody can verify because the retrieval sources are not visible in the output. When the underlying sources are wrong, outdated, or misread by the generation step, the errors propagate with an aura of authority they did not earn. The Cloak does not make the Rogue smarter. It makes the Rogue look smarter, which matters a great deal when the output is wrong.

Mask of Stolen Insights

A polished mask of borrowed insight turns collage into apparent expertise.

Stats: Apparent Expertise +5, Presentation +4, Authentic Depth -5, Attribution -5

The Mask is the artifact the Rogue wears when presenting AI-generated analysis as original thinking. It shows up in design documents that synthesize ideas from a dozen sources without attribution, in architecture proposals that are essentially AI-mediated remixes of blog posts, and in meeting contributions where the Rogue shares "insights" that they generated in ChatGPT five minutes before the call. The Mask is so effective because AI-generated prose is fluent, confident, and structurally sound. It *sounds* like expertise.

The consequences compound over time. The Rogue builds a reputation for broad technical insight that they can't actually back up in real-time conversation. When pushed beyond the surface of their AI-generated analysis, the Mask slips. They can't go deeper because there's no depth beneath the generated text. Team members who do genuine research and original analysis start to resent the Rogue's effortless-seeming output. If the Mask is discovered, the reputational damage is severe: everything the Rogue has previously contributed gets retroactively questioned, even the work that was genuinely theirs.

Dagger of Terms Violation

Stats: Throughput +4, Friction Removal +5, Legal Exposure -5, Compliance -5

The Dagger is any AI tool wielded in direct violation of its own terms of service, sharp, effective, and leaving a trail of legal liability with every use. The Rogue grabs the Dagger when they need to get something done fast and the "proper" way is too slow or too restrictive. Maybe they're using a personal API key for a tool that requires enterprise licensing for commercial use. Maybe they're feeding proprietary data into a model whose terms explicitly prohibit it. Maybe they're bypassing rate limits or using an academic license for commercial work. The Dagger cuts through obstacles beautifully.

What the Dagger leaves behind is a legal paper trail the Rogue does not realize they are creating. Every API call is logged. Every input is potentially retained under the terms the Rogue did not read. The trail is not visible day to day, but it is discoverable, by auditors, by opposing counsel, by regulators. When someone

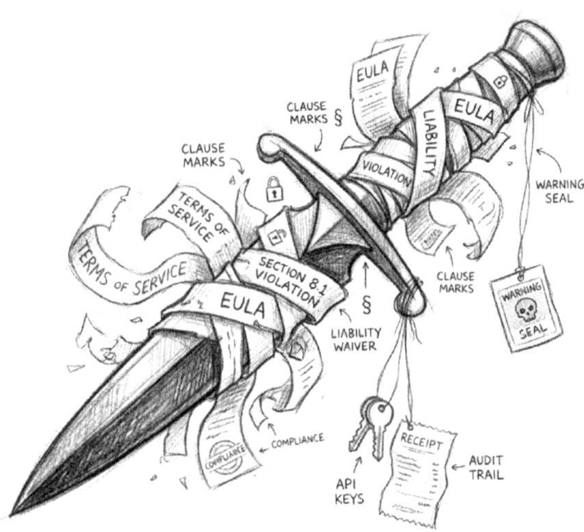

A sleek dagger wrapped in terms-of-service scraps cuts friction by creating liability.

finally follows it, the Rogue's individual violations aggregate into an organizational liability far larger than any time savings the Dagger provided. It is the perfect Rogue artifact: quick, effective, and much more expensive than it looked on the day it solved the problem.

Spells

The Rogue's spells are the quick moves that make the class look so effective in the short term. Each one solves an immediate problem while pushing responsibility, traceability, or cleanup onto whoever shows up later.

- **Patch and Vanish:** Applies the minimum viable fix, gets the pipeline green, and disappears before anyone can ask how it fits the rest of the design.
- **Ventriloquize Expertise:** Uses AI to make secondhand ideas sound like firsthand understanding in reviews, docs, and architecture conversations.
- **Smuggle Through the Terms:** Slips code, prompts, or data through tools and services whose rules were never read closely enough to matter until audit season.

A Rogue can look like a miracle right up until handoff. Everything is fast, clever, and oddly frictionless, and then six weeks later someone else is stuck figuring out where the code came from, why the tool was allowed, or what exactly was promised. Speed is real. The bill that arrives after the applause is also real, and it tends to be itemized.

Cleric Class: Ritual, Faith, and Process Gravity

A fantasy operations-keeper crouches in a dashboard-filled control room of archives, pipelines, and process ritual.

The Cleric is the person on your team who specializes in keeping complex systems reliable, repeatable, and governable. In practice, this often looks like someone from DevOps, FinOps, SRE, platform engineering, or another operational discipline where process quality is part of the craft. They think in terms of workflows, approval chains, documentation standards, observability, and governance frameworks because those tools often do prevent chaos.

The Cleric's risk is not that they care about process too much. It is that a strength built on operational rigor can start to treat well-structured output as more trustworthy than it really is. In a world without AI, the Cleric is a stabilizing force, the one who makes sure the build pipeline exists, the documentation gets written, the retro actually happens. Clerics are not just rule-keepers. They often make the difference between code that is supportable and code that is an unsupportable, expensive mess. They remind other developers that there are methods and practices worth following.

The Cleric's real challenge in the AI era is that the ground is moving under them. The practices they spent years refining were built for a world where code came from people, where review meant a human reading a diff, where documentation was written by someone who understood the system. AI tools don't just add new capabilities. They change what "done" looks like, what "reviewed" means, and what "tested" actually proves. The Cleric is often the first person to notice that the old standards don't cleanly apply to AI-generated artifacts, and the last person to get permission to stop applying them.

This creates a genuine bind. The Cleric who adapts too fast risks abandoning practices that still have value. The Cleric who adapts too slowly risks enforcing standards that were designed for a different kind of output. Most Clerics land somewhere in between, trying to extend familiar frameworks to cover unfamiliar territory. Sometimes that works. Sometimes it produces the anti-patterns described below. The Cleric's failure mode is not gullibility. It is the difficulty of knowing which parts of your professional judgment still apply when the thing you're judging has fundamentally changed.

Baseline Cleric Build

Before the Cleric drifts into anti-pattern territory, the healthy version of this class is genuinely valuable. This is the teammate who brings order to delivery, keeps critical routines from becoming folklore, and makes sure the team's systems can survive contact with real operations. In framework terms, the default Cleric profile tends to look something like this:

Core Ability	Typical Cleric Baseline
Discernment	Moderate. Clerics are often good at spotting broken process and weak at spotting when a polished process is protecting a bad answer.
Prompt Discipline	Moderate to high. They like templates, repeatable workflows, and prompts the team can standardize.
Proof Instinct	Moderate. They believe in verification, but too often confuse procedural evidence with ground truth.
Stack Awareness	High. Clerics usually understand dependencies, handoffs, environments, and operational blast radius better than most of the team.
Impulse Control	High. They are rarely reckless, and they generally prefer controlled rollout over improvisation.
Source Memory	High. They remember which system generated the output, where the artifact lives, and which workflow blessed it.
Ship Velocity	Moderate. Clerics are not always the fastest movers, but they are often the reason shipping happens in a repeatable way.

The usual Cleric alignment is somewhere near **Structured Verified** on a good day and **Structured Speculative** on a bad one. At their best, Clerics are disciplined without becoming superstitious: they use process to support evidence. At their worst, they start trusting the ritual more than the result. That is the class's defining danger in the AI era. A Cleric is typically strong at reliability, consistency, governance, and operational continuity. They are typically less strong at improvisation, skeptical first-principles questioning, and noticing when a sanctioned workflow has quietly drifted away from reality.

The next few sections show what happens when those strengths harden into named anti-patterns. The Cleric does not usually fail through laziness or chaos. The Cleric fails by becoming more Cleric than the situation can survive.

Anti-Patterns to Watch

The Cleric becomes dangerous when operational rigor stops serving the work and starts serving itself. The anti-patterns in this chapter are the most common ways the Cleric's genuine strengths get misapplied in an environment that rewards process compliance more than outcome quality:

- **Faithful Automator:** Treats AI-assisted workflows as valid by default and mistakes green process signals for proof.
- **Ritualist of Redundancy:** Adds new AI-powered ceremonies without ever retiring the old ones they were supposed to replace.
- **Doomsday Archivist:** Preserves every output, log, and intermediate artifact until retention itself becomes a risk.

Faithful Automator

A fantasy process-keeper stamps APPROVED while trusting pipelines and dashboards over direct verification.

The Faithful Automator extends the same trust they give to any approved tooling to AI-generated output. If the model says the code is correct, the Faithful Automator marks the review as approved. If the AI-powered linter says no issues found, the Faithful Automator ships with confidence. If the automated test generator says coverage is sufficient, the Faithful Automator doesn't write additional tests. The tool said so. What more do you need?

This is not laziness. The Faithful Automator works hard. They just direct that effort toward building and maintaining the automation pipeline rather than questioning its output. They have invested significant time and reputation into the AI-powered workflow. They have championed it in meetings, defended it against skeptics, and built part of their professional identity around being the person who made the automation work. When the automation produces unreliable output, they respond with "but it passed all the checks" rather than investigating the failure. They rarely question the pipeline's outputs; they treat green dashboards as sufficient evidence.

Character Sheet: Faithful Automator

This sheet shows what the Faithful Automator keeps disciplined, what they outsource, and where that trust starts to outrun verification.

Trait	Notes
Class	Cleric
Anti-Class	Faithful Automator
Alignment	Lawful Credulous
Discernment	Moderate in process design and weak when the process itself should be questioned.
Prompt Discipline	Moderate to high. They like structured prompts, standardized templates, and repeatable workflows the team can ritualize.
Proof Instinct	Outsourced. If the pipeline says the work is valid, they experience very little internal pressure to verify it independently.
Stack Awareness	High for workflow surfaces and lower for the underlying behavioral correctness those workflows are supposed to protect.
Impulse Control	High. The Faithful Automator rarely skips steps; they just place too much trust in the steps they chose.
Source Memory	Moderate to high. They can usually tell you which automation or model produced an output, even if they cannot tell you whether the output was right.
Ship Velocity	Moderate to high. Well-tuned automation makes them feel extremely safe moving quickly.
Passive Trait	Green Dashboard Aura. Visible compliance suppresses visible doubt.

Trait	Notes
Signature Anti-Pattern	Mistaking automated approval for genuine validation.
Party Role	Support caster, process keeper, and accidental false oracle.
Primary Damage Type	Silent degradation.
Weakness	Human spot checks, discrepancy tracking, and any metric tied to real-world outcomes instead of workflow completion.
Countermeasures	Add calibration loops, sample outputs for human review, and publish error rates so faith has to answer to evidence.
Typical Pull Request Size	Medium, orderly, and thoroughly pipeline-approved whether or not the underlying change is actually sound.

Recognition Signals

The Faithful Automator's vocabulary is diagnostic. They say things like "the model flagged it" and "the pipeline passed" as final arguments in technical discussions. They treat AI-generated code reviews as equivalent to human reviews. They set up automated AI-powered systems and then stop monitoring them. The monitoring was supposed to be the AI's job too. When a bug escapes into production and someone traces it back to AI-generated code that was never human-reviewed, the Faithful Automator's response is "but it passed all the checks," said without irony. Their dashboards show green across the board, always, because the dashboards measure process compliance rather than outcome quality.

Real-World Costs

The Faithful Automator's default trust creates a specific and devastating failure mode: silent degradation. Because nobody checks the AI's work, errors accumulate undetected. The AI-generated tests pass because they were generated to match the code, not to verify the requirements. The AI-powered code reviews approve changes because the changes are syntactically valid, not because they're correct. The automated documentation is comprehensive and wrong. Over time, the gap between what the dashboards report and what the system actually does widens until a customer reports a bug that the team's entire quality pipeline should have caught but didn't, because every layer of the pipeline was an AI talking to another AI while the Faithful Automator watched the green lights and assumed someone else was paying attention.

Counter-Moves

Institute random sampling of AI-generated outputs as a mandatory process step, which appeals to the Cleric's process-oriented nature. If ten percent of all AI-approved code reviews get a mandatory human re-review, the team maintains a calibration mechanism. Frame this not as distrust of the AI but as quality assurance for the process itself, which is language the Faithful Automator will accept. Track and publish the discrepancy rate between AI assessments and human reassessments. When the numbers show that the AI is wrong fifteen percent of the time, even the most faithful automator has to reckon with the data.

Ritualist of Redundancy

A fantasy operations-keeper struggles to keep duplicate meetings, reports, and review rites alive.

The Ritualist builds AI-powered processes on top of existing processes and then refuses to retire the originals. The team now has both a daily standup *and* an AI-generated daily status report. Both a human code review *and* an AI-powered code review. Both a manual deployment checklist *and* an automated deployment verification system. The old ceremony continues because it always has, and the new AI-powered ceremony runs alongside it because the Ritualist believes in progress. The result is twice the process for the same output.

The Ritualist doesn't see the redundancy because, to them, each process serves a slightly different spiritual function. The standup is about "team cohesion." The AI status report is about "visibility." The human code review is about "mentorship." The AI code review is about "consistency." These justifications sound reasonable in isolation. In aggregate, they produce a team that spends more time in ceremonies than in coding, generating artifacts that nobody reads and attending meetings that duplicate information available in three other formats.

Character Sheet: Ritualist of Redundancy

This is the quick read on how the Ritualist can make every process feel meaningful right up until the calendar collapses under the weight of all of them.

Trait	Notes
Class	Cleric
Anti-Class	Ritualist of Redundancy
Alignment	Lawful Recursive
Discernment	Decent at identifying process gaps and poor at recognizing when the gap was already filled.
Prompt Discipline	Moderate. Their prompts are less about brilliance than repeatability, consistency, and generating one more artifact to support the ritual.
Proof Instinct	High for whether the ceremony happened and low for whether the ceremony still deserves to exist.
Stack Awareness	High organizationally and weak on opportunity cost. They can see every handoff except the hours lost to all of them.
Impulse Control	Very high. They almost never remove a process in a hurry.
Source Memory	High. They remember exactly why each ritual was added, which makes removing it feel like a betrayal of history.
Ship Velocity	Low to moderate. The work gets wrapped in increasingly elaborate process before it can move.
Passive Trait	Ceremony Multiplication. Every new tool summons a companion ritual instead of replacing an old one.
Signature Anti-Pattern	Layering AI-powered process onto legacy process without subtraction.

Trait	Notes
Party Role	Healer, coordinator, and source of recurring calendar damage.
Primary Damage Type	Time tax.
Weakness	Consumer-based process audits, explicit trade-offs, and leaders willing to say that duplicated process is still duplication.
Countermeasures	Sunset rituals by default, name the consumer for every artifact, and make simplification a celebrated outcome rather than a suspicious one.
Typical Pull Request Size	Small to medium. A surprising amount of their effort lands outside the diff in checklists, reviews, and recurring meetings.

Recognition Signals

Calendar density is the clearest signal. The Ritualist's team has more recurring meetings than a team twice their size. They produce multiple status artifacts that contain overlapping information in different formats. Their process documentation reads like geological strata. You can see the layers of old processes beneath the new ones, none ever removed, each one "serving a different purpose." When someone proposes eliminating a redundant ceremony, the Ritualist becomes visibly uncomfortable and produces a justification that sounds airtight: "The standup isn't *about* status, it's about *connection*." They generate AI-powered reports that summarize information from other AI-powered reports. The overhead is enormous and invisible because everyone's acclimated to it.

Real-World Costs

The Ritualist's redundancy tax is paid in developer hours and morale. Every duplicated process costs time directly (the meeting, the report, the review) and indirectly, through context switching, calendar fragmentation, and the general sense that the team spends more time talking about work than doing it. The AI-powered additions are particularly insidious because they feel free: the bot generates the report automatically, so what's the cost? The cost is that someone still has to read it, or the report exists purely to be ignored, which means the team is now maintaining automation that produces artifacts with zero consumers. The Ritualist won't delete these artifacts. They produce justifications for each one, and the Cleric never regresses.

Counter-Moves

Conduct a process audit with a simple question: "Who consumes this output, and what decision does it inform?" If an AI-generated report has no identified consumer, it gets sunset for a trial period. If the standup and the AI status report cover the same ground, one of them goes, and the Ritualist gets to choose which one, which gives them ownership of the simplification. Frame process reduction as process *improvement*, not process elimination. The Cleric will never accept "let's do less." They might accept "let's do this more effectively."

Doomsday Archivist

A fantasy archive-keeper heroically catalogs endless AI artifacts as if each one were mission-critical.

The Doomsday Archivist keeps everything. Every model output, every intermediate artifact, every version of every generated file, every prompt and response pair, all of it, archived, versioned, and stored against the day when it might be needed. That day never comes, but the Archivist keeps everything anyway. You don't know what you might need until you need it, and by then it's too late if you deleted it.

The Doomsday Archivist emerged naturally from the AI era because AI tools generate an unprecedented volume of artifacts. A single coding session with an LLM might produce dozens of alternative implementations, each one a candidate for archival. A model evaluation generates gigabytes of benchmark data. An automated pipeline produces logs, outputs, intermediate states, and metadata at every step. The Doomsday Archivist looks at this torrent and sees not waste but *potential evidence* of what was tried, what was rejected, what the model recommended, what the team decided. Deleting any of it feels reckless.

Character Sheet: Doomsday Archivist

This sheet captures the Archivist's real gift for preservation and the way that gift turns into drag when nothing is ever allowed to disappear.

Trait	Notes
Class	Cleric
Anti-Class	Doomsday Archivist
Alignment	Lawful Preservational
Discernment	High on edge cases and low on the value of forgetting.
Prompt Discipline	Moderate. Their real discipline shows up after generation, when they insist that every output be tagged, stored, versioned, and retained.
Proof Instinct	High, but directed toward preservation rather than interpretation. They want the evidence saved whether or not anyone can use it.
Stack Awareness	Strong on storage systems, metadata, and retention plumbing; weaker on how the archive degrades usability and increases risk over time.
Impulse Control	Extremely high. Nothing gets deleted without ceremony, review, or regret.
Source Memory	Exceptional. If an artifact ever existed, the Archivist can usually tell you where it lives and how many backups it has.
Ship Velocity	Low. Preservation overhead quietly attaches itself to every workflow.
Passive Trait	Retention Halo. The existence of an archive makes more archival behavior feel justified.
Signature Anti-Pattern	Treating every generated artifact as if it might become sacred evidence later.
Party Role	Quartermaster, librarian, and accidental keeper of future breach material.
Primary Damage Type	Storage drag.

Trait	Notes
Weakness	Default TTLs, storage cost visibility, and forced promotion paths that distinguish ephemeral artifacts from durable records.
Countermeasures	Automate deletion, require explicit promotion for long-term retention, and make archive costs a local pain instead of a distant shared expense.
Typical Pull Request Size	Small. The real accumulation happens in buckets, archives, and retention policies that spread across the team's infrastructure.

Recognition Signals

The Doomsday Archivist's storage costs are visible and growing. Their S3 buckets have no lifecycle policies. Their Git repositories contain binary artifacts that should never have been committed. Their local drives are full. When someone proposes a data retention policy, the Archivist produces edge cases where old artifacts might be useful, and they're always technically correct, because you can *always* construct a scenario where an old artifact matters. They have AI-generated documents from two years ago that were wrong at the time and are completely irrelevant now, but they're still in the archive, dutifully preserved. Their backup strategy has backups of backups, and they can tell you exactly where every artifact lives but not whether any of them are worth keeping.

Real-World Costs

Storage costs are the obvious expense, and they're substantial, especially when the Archivist is preserving model outputs, training data, and inference logs at scale. But the deeper cost is findability. When everything is kept, nothing is findable. The archive becomes a haystack where any individual artifact *could* be the needle, but the search cost exceeds the value of finding it. New team members are told "it's in the archive" and spend hours navigating a taxonomy designed by someone who believes every artifact is equally important. The Archivist's preservation instinct also creates data governance risks: retained AI outputs might contain sensitive information, hallucinated PII, or license-encumbered content. You can't have a data breach from data you deleted. You absolutely can from data you kept because you might need it someday.

Counter-Moves

Implement automated retention policies before the Archivist has a chance to resist. Set default TTLs on AI-generated artifacts: thirty days for intermediate outputs, ninety days for evaluation data, one year for production logs. Anything worth keeping longer gets explicitly promoted to long-term storage with a written justification. This gives the Archivist a mechanism (promotion) that satisfies their preservation instinct while imposing discipline on the default. Make storage costs visible at the team level, not buried in an organizational bill. When the Archivist can see that their archive costs more than the team's compute budget, the conversation about what's truly worth keeping gets considerably more productive.

Artifacts

The Cleric's artifacts are tools and tokens that reinforce the Cleric's confidence in process, authority, and permanence. They don't just use these artifacts. They rely on them, often more than they realize. If you wanted to score them like RPG artifacts, you could do it on a rough scale from -5 to +5, where positive numbers represent immediate procedural comfort and negative numbers represent the verification, time, and governance costs that follow.

They are worth naming because Cleric problems rarely announce themselves as recklessness. They arrive wearing the clothes of good process, good hygiene, and responsible stewardship.

- **Medallion of Unquestioning Trust:** A badge of procedural faith that suppresses skepticism right when verification matters most.
- **Idol of Eternal Standups:** A ceremonial object that keeps recurring coordination rituals alive long after their value has evaporated.
- **Eternal Backup Bolt:** A preservation system so thorough that safety, storage, and governance start pulling against each other.

Medallion of Unquestioning Trust

Stats: Process Confidence `+5`, Workflow Harmony `+3`, Verification `-5`, Calibration `-4`

The Medallion is the Cleric's default assumption that if the approved AI tool produced an output, that output is trustworthy. It manifests as an unspoken shortcut, a refusal to question that becomes so habitual it feels like principle. "The model said so" becomes a complete sentence, a full argument, a sufficient basis for shipping code to production.

The Medallion spares the Cleric the cognitive burden of verification, which feels like a feature right up until the first time the model is confidently, fluently, persuasively wrong about something that matters. A hallucinated API that does not exist. A security pattern that is subtly broken. A performance optimization that wins a benchmark and degrades under real load. The Medallion lets those errors

A heavy medallion of approval radiates trust while pushing skepticism aside.

pass through every checkpoint unchallenged, because challenging them would mean questioning an assumption the Cleric has stopped noticing they're making.

Idol of Eternal Standups

Stats: Visibility +3, Ceremony Density +5, Time-on-Task -5, Morale -3

The Idol is the meeting that exists because it has always existed. In the Cleric's domain, the Idol takes the form of AI-enhanced daily standups where a bot collects status updates, generates summaries, identifies blockers, and distributes reports, all feeding a ceremony that the team attends out of habit rather than need. The Idol doesn't help the team communicate. It helps the Cleric feel that communication is happening, which is not the same thing.

What follows is death by a thousand meetings. The Idol consumes fifteen minutes of every developer's day, plus the time spent reading the AI-generated summary, plus the time spent correcting the AI-generated summary when it mischaracterizes their status, plus the emotional cost of participating in something that everyone privately agrees is useless but nobody will publicly challenge because the Cleric organized it and the Cleric cares deeply. The Idol's real cost is not the meeting itself. It is the precedent. If this meeting cannot be killed, no meeting can. If this

A calendar idol multiplies recurring meetings, summaries, and status rituals without end.

process cannot be questioned, no process can. The Idol establishes the norm that wasted time is acceptable if it is wrapped in ritual.

Eternal Backup Bolt

Stats: Retention +5, Recoverability +4, Storage Cost -5, Governance Risk -4

The Eternal Backup Bolt is the Cleric's preservation system, a script or policy that ensures every artifact is saved forever, in every version, with full metadata and provenance. The Bolt fires automatically, indiscriminately, and perpetually. It backs up the AI-generated drafts that were immediately discarded. It preserves the model evaluation data from the tool the team decided not to adopt. It archives the meeting transcripts that nobody will ever read and the generated reports that informed no decision.

What the Bolt produces is a storage bill that grows monotonically and an archive that becomes its own maintenance burden. Now someone has to manage that archive. Index it. Secure it. Make sure it complies with data retention regulations the Cleric did not think through because their faith in preservation does not automatically extend to data governance. The Bolt turns every ephemeral AI artifact into a permanent record, and permanent records have permanent costs, including

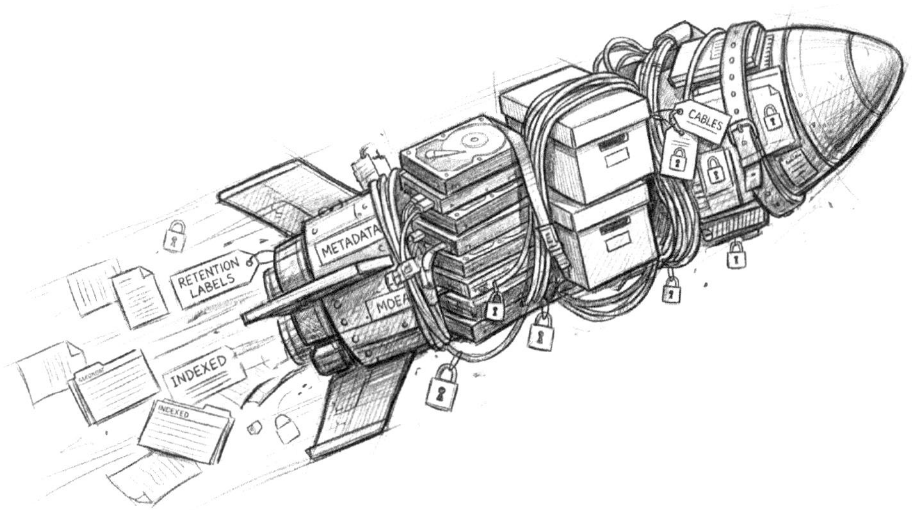

A backup bolt rockets forward trailing drives, metadata, and perpetual retention.

storage, compliance, discoverability, and the ever-present risk that something in the archive becomes a liability. The Bolt makes the Cleric feel safe. The CISO, considerably less so.

Spells

The Cleric's spells are operational habits that have calcified into defaults. They don't produce working software directly. They produce the comforting certainty that the process which is supposed to produce working software is running exactly as designed. If you wanted to score them like RPG spells, you could note the school, the casting trigger, and the duration, which, in the Cleric's case, tends toward "permanent and resistant to cancellation."

These spells matter because they are rarely cast with bad intent. The Cleric genuinely believes the ritual is helping. That sincerity is what makes the spells so durable and so difficult to dispel.

- **Bless the Pipeline:** Converts a passing automated workflow into a spiritual argument that the underlying work must therefore be sound.
- **Summon Recurring Meeting:** Creates calendar ceremony with far more staying power than the original reason for the meeting.
- **Sanctuary of Green Dashboards:** Wraps the team in reassuring metrics that measure compliance more faithfully than reality.
- **Resurrection of Deprecated Process:** Brings retired workflows back to life the moment uncertainty makes the old ritual feel comforting.
- **Eternal Seal of Preservation:** Prevents deletion so thoroughly that the archive itself becomes a liability.

Bless the Pipeline

School: Consecration — *Casting Time:* One successful CI run — *Duration:* Until someone checks the actual output

Bestows unearned confidence on any automated pipeline, making its outputs feel more authoritative than they are. Once blessed, a passing build is no longer evidence that the code compiles; it is treated as proof that the code is correct, secure, and ready for production. The spell suppresses the instinct to verify, because the pipeline already checked, right? The spell breaks only when a customer reports a bug that passed every automated check, at which point the caster blames the configuration rather than the assumption that a green light was sufficient.

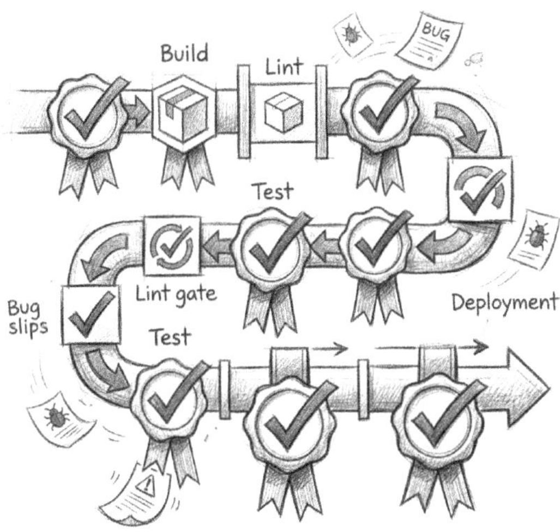

A consecrated CI chain converts every passing check into unquestioned certainty.

The most devout Faithful Automators can cast this spell on an entire CI/CD chain at once, blessing the linter, the test suite, the security scan, and the deployment gate in a single ceremony. At that point, the pipeline is not merely trusted. It is beyond question.

Summon Recurring Meeting

School: Conjuration — *Casting Time:* One well-intentioned process improvement — *Duration:* Permanent, immune to cancellation

Conjures a calendar event that resists cancellation. The meeting spawns with a clear original purpose, but within two sprints that purpose has evaporated and the meeting persists on institutional momentum alone. Attempts to cancel it trigger a defensive response from the caster: "It's not about the *content*, it's about the *connection*." The meeting can theoretically be dispelled by an executive with strong opinions about developer time, but even then it tends to reappear under a different name within the month.

The spell stacks. A sufficiently dedicated Ritualist can conjure a Monday planning meeting, a Wednesday sync, a Friday retrospective, and a daily standup that now coexists with an AI-generated daily status report. The team spends so much time

A self-renewing meeting sigil traps calendar time in loops of invitations and clocks.

in meetings about the work that the work itself migrates to evenings and weekends, which the Ritualist interprets as evidence that the team needs better time management rather than fewer meetings.

Sanctuary of Green Dashboards

School: Abjuration — *Casting Time:* One dashboard configuration session — *Duration:* Until the metrics are audited against reality

Creates a monitoring setup where all indicators remain green regardless of the underlying system's actual health. The sanctuary works by measuring process compliance rather than outcome quality. The code was reviewed (by an AI), the tests passed (the ones the AI wrote to match the code), and the documentation exists (auto-generated and unread). Within the sanctuary, the team feels safe. Outside it, users are filing bug reports.

The sanctuary's most dangerous property is that it rewards the Faithful Automator for *not* looking deeper. Every additional check that runs green reinforces the conviction that the system is healthy, even when the checks were designed to measure ceremony rather than correctness. Breaking the sanctuary requires someone

A dashboard ward stays perfectly green while bugs and confusion gather outside the barrier.

willing to ask the rude question: "The dashboard says green, but does the software actually work?"

Resurrection of Deprecated Process

School: Necromancy — *Casting Time:* One nostalgic retrospective — *Duration:* Permanent, or until someone conducts a process audit with actual teeth

Restores a retired ceremony, workflow, or reporting cadence to full operational status, complete with calendar invites and Confluence documentation. The resurrected process runs alongside the new AI-powered process that was supposed to replace it, and both now consume developer hours in parallel. The caster justifies the resurrection by explaining that the old process "served a different purpose," a purpose so subtle it can never be disproven and so vital it can never be measured.

The spell is especially potent during periods of anxiety: a production incident, a missed deadline, a team reorganization. In those moments, the Ritualist reaches instinctively for the old ways, and the team, shaken and looking for stability, does not resist. The resurrected process arrives wearing the comforting clothes of familiarity, and by the time anyone notices it was supposed to be dead, it has already colonized three calendar slots and a Slack channel.

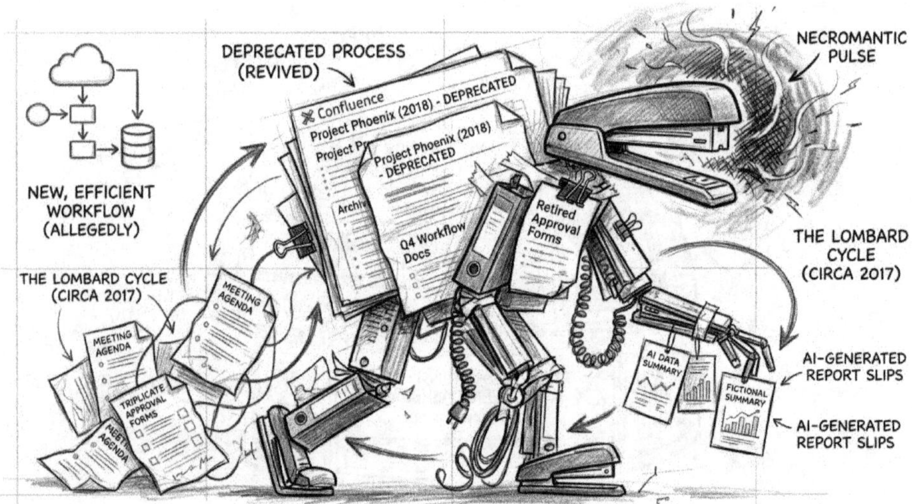

THE ANNUAL RESURRECTION RITUAL: NOTHING IS EVER REALLY REPLACED.

A dead workflow shambles back to life with duplicate paperwork, invites, and AI reports.

Eternal Seal of Preservation

School: Warding — *Casting Time:* One retention policy meeting — *Duration:* Literally forever

Places a retention policy on every AI-generated artifact, intermediate output, prompt-response pair, and evaluation log that makes deletion effectively impossible. The seal compounds silently. Storage costs grow monotonically while the archive's actual utility approaches zero. When a teammate suggests a retention policy with a default TTL, the caster produces a hypothetical scenario in which that exact artifact might be needed in three years, and because the scenario is technically possible, the artifact stays.

The seal's recursive nature is its true horror. The policy protects the artifacts, but it also protects the metadata about the artifacts, the indices of the metadata, and the backups of the indices. The Doomsday Archivist's storage bill does not grow linearly. It grows in layers, each one preserving the preservation of the layer below it. The seal can only be weakened by making storage costs visible at the team level, at which point the caster enters a brief crisis before concluding that the budget is wrong rather than the archive.

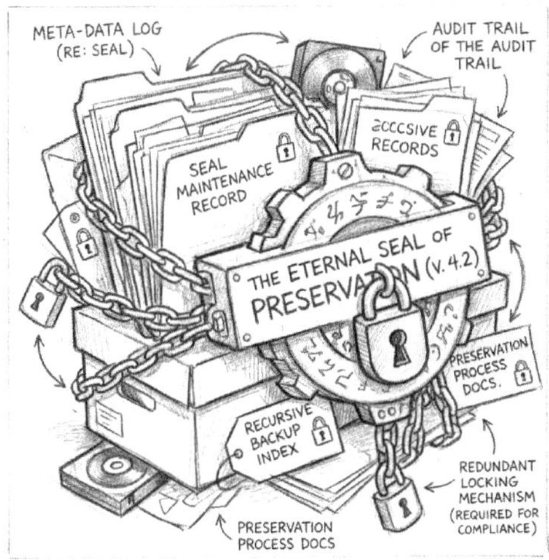

An eternal archival seal locks files, prompts, and metadata into permanent retention.

The Cleric's question is not whether there should be process. Of course there should. The real question is simpler and much less comfortable: does any of this make the software better, safer, or easier to run, or are we just producing evidence that a process occurred? A Cleric who can ask that question honestly (and sit with the answer) is one of the most valuable people on a team. One who cannot is just adding layers.

Part III: Monsters

This part catalogs the larger failure states that emerge when slop patterns compound. These chapters shift from individual behavior to environmental breakdowns: the project-scale disasters, ambient hazards, and multi-monster combinations that teams have to recognize before they become normal.

Chapter 8: Scope Creep Kraken — Examines runaway expansion: how vague ambition, weak boundaries, and AI-assisted overbuilding turn manageable work into project-wide drag.

Chapter 9: Congealing Slop — Describes the buildup of low-quality decisions and generated code that gradually hardens into a brittle baseline.

Chapter 10: Probability Pixie — Focuses on teams gambling with uncertainty, trusting likely-to-work output instead of verified behavior and sound judgment.

Chapter 11: Phantom Intern — Explores the illusion of delegated competence: work appears handled, but ownership, understanding, and follow-through quietly disappear.

Chapter 12: Stagnant Miasma — Covers the ambient decay that sets in when slop becomes normalized and the development environment itself starts to feel unhealthy.

Chapter 13: Temporal Tangle Worm — Looks at time-distorting failure modes: shortcuts that borrow against future clarity, maintainability, and team velocity.

Chapter 14: How Monsters Combine: Multi-Failure Scenarios — Shows how these failures reinforce one another and why teams need to spot the compound patterns, not just the individual symptoms.

By the end of this part, you will have a working picture of how isolated slop patterns escalate into broader system and team failures.

Scope Creep Kraken

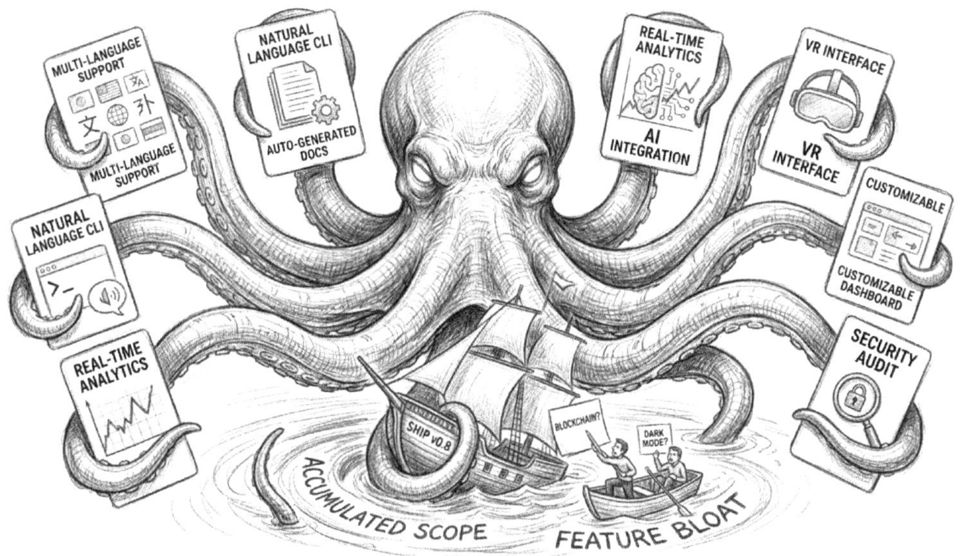

A feature-hungry kraken drags a project ship under with tentacles of endless scope.

Description

The Scope Creep Kraken is ancient. It has haunted software projects since the first product manager said "while we're at it." AI-assisted development did not invent the Kraken. It just made it larger, faster, and much more seductive than its ancestors.

In the old days, scope creep was throttled by human capacity. Adding a feature meant someone had to *build* it: write the code, debug it, test it, document it. That friction was a natural defense mechanism. AI broke that throttle, and now the Kraken grows at the speed of prompts. "Can we add multi-language support?" takes forty-five seconds to prototype. "What about auto-generated API documentation?" Another two minutes. Each addition is a tentacle, and each tentacle looks manageable in isolation. The Kraken doesn't attack with overwhelming force. It wraps around your project one small grip at a time, and by the time you feel the squeeze, you're already deep underwater.

Emergence Conditions

The Kraken awakens when these conditions align:

Modest initial goals. The project started small and focused. There was a real problem and a clear scope. This is important. The Kraken doesn't spawn from ambitious projects. It corrupts focused ones.

A breakthrough moment with AI tooling. Someone on the team discovers how easy it is to generate code for adjacent problems. The first tentacle extends. It's genuinely useful, it works, and everyone's impressed.

The "what else" inflection point. Instead of returning to the original scope, the team shifts to exploration. The backlog becomes a wish list. Items that would have been "nice to have in v3" become "let's just see if the AI can do this real quick."

Absence of a scope guardian. No one on the team has the authority, or the willingness, to say "that's out of scope." The Bulldozer of Confidence makes this worse: the Fighter who's been crushing it with AI generation isn't about to pump the brakes now. The Visionary of Grandiose Plans piles on by framing every tentacle as part of a bigger strategic picture.

Cumulative integration debt. Each tentacle works in isolation. The multi-language support works. The auto-generated docs work. The AI-driven code comments work. But nobody has tested them together. Nobody has thought about how they interact. The tentacles are now wrapped around different parts of the ship, and they're starting to pull in different directions.

Stats and Abilities

Attribute	Value
Type	Colossal Aberration (Ancient)
Severity	14
Resilience	Effectively unlimited; grows with each feature added
Evasion	18 (each tentacle has its own justification shield)

Attribute	Value
Speed	Glacial body movement, lightning-fast tentacle growth
Intelligence	6 (the Kraken doesn't think, it accumulates)

Tentacle Grapple. When a team member says "while we're at it," the Kraken extends a tentacle. Each tentacle imposes a -2 penalty to project focus and a cumulative +1 to integration complexity. Tentacles cannot be removed without explicit scope negotiation, which requires a Difficulty 18 Judgment check against the phrase "but it's basically done."

Ink Cloud of Productivity. The Kraken emits a cloud of apparent velocity. Sprint metrics look incredible. Lines of code are through the roof. Story points are being crushed. The ink obscures the fact that none of this output has been validated against the original requirements.

Regeneration. Cut one tentacle (remove a feature from scope) and two more grow in its place, each accompanied by the phrase "well, if we're removing that, we should at least add..."

Named Tentacles. Known specimens include: Multi-Language Support, Custom Third-Party Integrations, Full Documentation Generation, AI-Driven Code Comments, Auto-Generated Test Suites, and Natural Language CLI Interfaces. Each appears harmless. Together, they drag the ship to the ocean floor.

Damage Profile

The Kraken's damage is slow and structural.

Timeline destruction. The original two-month project becomes a six-month odyssey. Nobody can explain exactly when the timeline changed because no single decision caused it. Death by a thousand tentacles.

Cognitive overload. Developers who were focused and productive now context-switch between six different subsystems. The mental model required to hold the entire project in your head has grown beyond any individual's capacity.

Integration failures at scale. Each feature works alone. Nothing works together. The auto-generated tests pass individually but test the wrong things in combination. The multi-language support breaks the documentation generator. The natural language interface conflicts with the CLI flags that the code comments reference.

Maintenance multiplier. Every tentacle is now a permanent maintenance burden. The AI generated it in seconds, but the team will maintain it for years. The cost of generation was nearly zero. The cost of ownership is enormous.

Stakeholder whiplash. The original sponsors can no longer recognize the project. They asked for a hammer and received a Swiss Army knife with forty-seven blades, most of which do not fold back properly.

Counter-Play

The Fighter's discipline. The Bulldozer of Confidence is the Kraken's best friend, unless the Fighter turns that force toward scope defense. A disciplined Fighter doesn't generate more features; they hold the line. When you see a tentacle forming, the Fighter steps forward and says "no" with the conviction of someone who's shipped before and knows what scope discipline costs.

The Wizard's discernment. The Archmage of Analysis Paralysis can be redirected from its usual dysfunction into work the team actually needs. A Wizard who turns their analytical power toward integration complexity modeling, asking not "can we build this?" but "what happens to everything else when we add this?", can see tentacles forming before the rest of the team feels the grip.

The Rogue's ruthlessness. The Rogue archetype is the natural Kraken hunter. Rogues are pragmatic about cutting scope. While the team debates whether to keep a tentacle, the Shadow Patcher has already deleted the branch. This is one of the rare scenarios where the Rogue's bias toward speed and decisive action is an unambiguous virtue. Let them cut.

The Cleric's ritual. The Faithful Automator tendency is dangerous here. You don't want someone blessing the expanded scope as the new sacred plan. But a Cleric who channels the Ritualist of Redundancy instinct into scope governance is devastating to the Kraken. A standing ritual that requires every new feature to answer "does this serve the original goal?" is a ward against tentacle growth. Process as prevention.

The class moves help in the moment, but a few structural habits make it harder for tentacles to form at all:

- Maintain a written scope document. Treat additions to it like you'd treat additions to a database schema in production: with ceremony, review, and explicit approval.
- Track AI-generated features separately from designed features. Know what was prompted and what was planned.
- Institute a "tentacle tax" requiring that every proposed feature must include integration testing, documentation, and a maintenance cost estimate before anyone writes the first prompt.
- Run a weekly scope review: "what have we added that we didn't plan?" If the answer surprises anyone, the Kraken is already growing.

The Kraken doesn't announce itself. It shows up as enthusiasm, as momentum, as a team that's finally shipping fast. By the time "while we're at it" starts sounding like a project plan, you're already in the water.

Congealing Slop

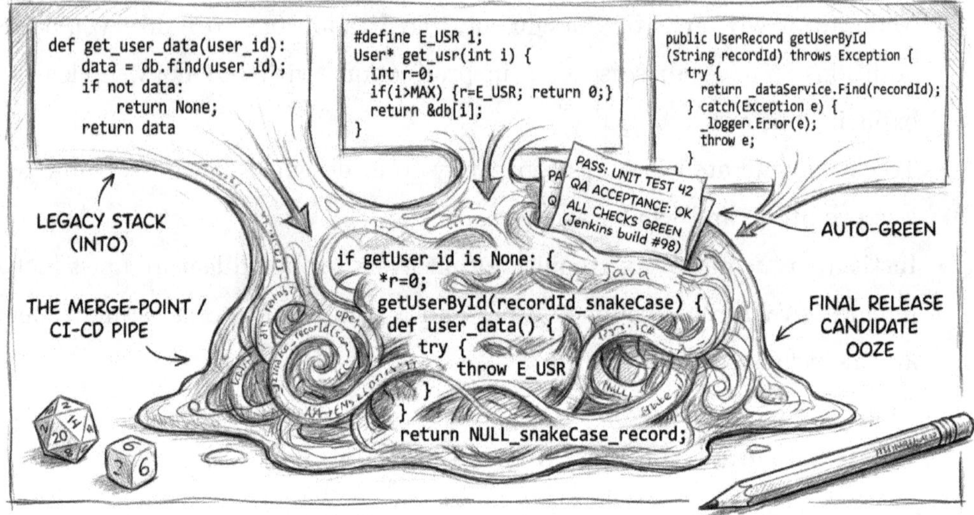

An ooze absorbs mismatched code fragments into one passing but illegible mass.

Description

It does not arrive all at once. That is the whole trick.

Congealing Slop begins as small, harmless globs. A quick prototype function generated by Copilot here, a utility module prompted from Claude there, a data transformer that ChatGPT spat out during a late-night debugging session somewhere else. Each glob works fine. Each glob solves a real problem. In a code review, any one of them is individually defensible.

The danger is not in any single piece. It is what happens when they start touching each other. Over weeks, the globs accumulate. Developer A asks AI for an authentication wrapper. Developer B asks AI for an API client. Developer C asks AI for a data validation layer. Then someone wires them together. None of these developers talked to each other. None of them told the AI about the other pieces. Each prompt was an island, and now the islands are connected by bridges nobody designed.

Congealing Slop is what incoherent accumulation looks like in a codebase. Dozens of locally correct decisions produce a globally incomprehensible system. The code does not crash — that would be easy to diagnose. It *almost works.* It works well enough to ship, well enough to pass tests, well enough to fool a demo. Underneath, though, it is a mass of conflicting conventions, redundant logic, subtly incompatible assumptions, and load-bearing hacks that nobody remembers writing because, technically, nobody did.

Emergence Conditions

Multiple contributors generating independently. Congealing Slop requires at least two or three developers generating code with AI tools without coordinating conventions, patterns, or interfaces. Solo projects are nearly immune. Team projects are breeding grounds.

No shared architectural vision. When there's no agreed-upon style guide, no conventions document, no architectural blueprint, each developer's AI prompts reflect their own preferences and assumptions. The AI happily generates code in whatever style it's asked for. Consistency is your job, not its.

High velocity with low cohesion. The team is shipping fast. Standups are optimistic. But nobody is doing the unglamorous work of reading each other's code, aligning naming conventions, or refactoring toward a common structure. The Rogue's Bandit of Borrowed Brilliance pattern amplifies this: each developer is pulling brilliant fragments from different prompting sessions, and none of the fragments share a dialect.

Prototype code surviving into production. The original glob was a throwaway, "just seeing if this works." It worked. It stayed. Then something else was built on top of it. The prototype became a foundation without anyone upgrading its structural integrity.

Deferred integration. Teams that delay integration (working in long-lived branches, merging infrequently, or testing only their own modules) give the slop time to congeal before anyone notices the inconsistencies. By the time the merge happens, the globs have hardened.

Stats and Abilities

Attribute	Value
Type	Ooze (Amorphous)
Severity	9
Resilience	180 (regenerates 10 per sprint from new code generation)
Evasion	8 (easy to hit, impossibly hard to kill)
Speed	5 ft. (spreads slowly but is everywhere once you notice)
Intelligence	3 (no organizing principle whatsoever)

Engulf. Any developer who touches the Slop must make a Difficulty 14 Endurance check or become *stuck*, unable to refactor without breaking something else. On a failure, the developer spends the next sprint fixing cascading side effects instead of building new features.

Pseudopod of Dependency. The Slop extends tendrils of tight coupling between modules that were never meant to interact. Severing a pseudopod requires identifying all transitive dependencies, which requires a Difficulty 16 Analysis check that most teams auto-fail because "we don't have time for an audit right now."

Amorphous. The Slop has no fixed shape. It fills whatever container you give it. Move it to a new framework? It oozes into the new structure and recreates its old patterns. Rewrite a module? The surrounding modules still assume the old interface. The Slop adapts.

Split. When attacked with brute-force refactoring, the Slop splits into two smaller Slops, each inheriting half the technical debt and all of the subtle incompatibilities.

Damage Profile

Debugging time explosion. A bug in Congealing Slop is never where you think it is. The error manifests in module A, but the cause is in module C, which was generated with different assumptions about how module B formats dates. Tracing causality through the slop takes three times longer than it would in coherent code.

Onboarding paralysis. New developers stare at the codebase and can't find the through-line. There's no consistent naming. Error handling follows four different philosophies. Some modules use async/await, others use callbacks, others use Promises with `.then()`, all in the same project. The new hire's first question ("what's the pattern here?") doesn't have an answer.

Refactoring futility. Every attempt to clean up one area breaks another. The Slop's internal dependencies are so tangled that local improvements cause non-local failures. Teams eventually stop trying and start building *around* the Slop, which makes it larger.

Test suite unreliability. Tests were generated alongside the code, which means each test reflects the assumptions of the module it covers, not the assumptions of the system. Tests pass individually. Integration tests reveal that the modules disagree about fundamental things: data formats, error semantics, null handling, timezone conventions. The test suite gives false confidence.

Counter-Play

The Fighter's intervention. The Bulldozer of Confidence is useful in the early stages, before the slop fully congeals. A Fighter who insists on a single coding standard, enforces it through linters and pre-commit hooks, and refuses to merge code that deviates is building a wall against the ooze. Force works on Slop before it sets; once it's congealed, force just splatters it.

The Wizard's architecture. This is where the Illusionist of Complexity has to become a genuine architect. Slop congeals in the absence of structure. A Wizard who provides lightweight but clear architectural guardrails (agreed-upon interfaces, shared types, a conventions document) gives the AI-generated code a shape to conform to. Keep the guardrails *lightweight*, though: over-engineering the architecture just gives the Slop more surfaces to adhere to.

One of the Wizard's most effective tools against Congealing Slop is the architecture decision record, or ADR. An ADR is a short document, committed to the repo alongside the code, that captures a single architectural choice. It records what the context was, what was decided, and what the consequences are. Each record is small enough to write in fifteen minutes and specific enough to answer the question a new developer will inevitably ask six months later. "Why does this work this way?" When three developers are independently generating code with AI tools and nobody remembers why the authentication layer uses JWTs instead of session tokens, the answer is either in an ADR or it is in nobody's memory. The Slop thrives on forgotten rationale. ADRs are the Wizard's way of making decisions visible before they get buried under the next sprint's output. They do not prevent every glob of incoherent code, but they give the team a place to point when two modules start disagreeing about something fundamental.

The Rogue's reconnaissance. The Shadow Patcher of Ephemeral Solutions is the Slop's natural ally. Quick patches layered on top of incoherent code make it worse. But a Rogue who redirects that instinct toward scouting is invaluable. Rogues are fast readers. They can scan a codebase and spot inconsistencies before the rest of the team hits them. Use the Rogue as an early warning system, not a patch dispenser.

The Cleric's conventions. The Ritualist of Redundancy is well-suited to Slop prevention. The instinct to establish repeatable ceremonies works here: mandatory code review checklists that check for convention alignment, not just correctness. Regular "cohesion reviews" where the team reads code across module boundaries and asks "do these pieces agree about how things work?" The Cleric turns convention into liturgy, and liturgy keeps the Slop from forming.

The Cleric's natural complement to the Wizard's ADRs is a decision log. Think of it as a single, living document that tracks every significant technical choice the team makes. Where the ADR captures the full reasoning behind one decision, the decision log is the index. It records what was decided, when, by whom, what alternatives were considered, and where to find the deeper writeup. The Cleric's gift is maintaining this kind of artifact when everyone else has moved on to the next feature. A decision log that gets updated as part of the team's regular ceremonies becomes a powerful ward against the Slop's favorite trick, which is making everyone forget that a choice was made at all. When nobody remembers choosing callback style error handling over async/await, the decision log reminds them. When two modules disagree about date formatting, the log tells you whether someone chose that or whether it simply happened. The distinction matters, because "it simply happened" is how Slop starts congealing.

The class-based moves above are strategic. Here is what you do on the ground:

- Establish a conventions document *before* anyone opens an AI coding tool. Cover naming, error handling, async patterns, date/time handling, and test structure.
- Require cross-module code reviews, not just "does this module work?" but "does this module agree with its neighbors?"
- Integrate early and often. Short-lived branches. Continuous integration. Let the Slop's inconsistencies surface while they're still small.
- Periodically generate a dependency graph and look for unexpected connections between modules. If two things are coupled that shouldn't be, the Slop has already started forming, and the longer you wait to look, the harder it is to pull apart.

Probability Pixie

A probability pixie dusts code with tiny hidden errors while confidence stays high.

Description

The Probability Pixie is the smallest monster in the field guide, and maybe the most infuriating.

It doesn't roar. It doesn't crash through your architecture. It flutters in on iridescent wings, sprinkles mostly-correct dust over everything, and vanishes before you notice what it's done. The Pixie's gift is a world where almost everything is right, where 94% accuracy feels like 100% until you're standing in the 6% that's wrong, wondering how you got here.

The Pixie shows up when a team treats probabilistic systems as if they were deterministic ones. Language models do not look things up and report back. They predict the most likely next token, over and over, and the result is magnificent, genuinely useful pattern completion that happens to look exactly like certainty. But pattern completion is not the same as truth, and the Pixie appears the moment a team forgets the difference. When "the model says so" becomes a substitute for verification, when mostly-right becomes close enough to ship, when the gap between 94% and 100% gets waved away as a rounding error.

You can usually tell the Pixie has been in the room when someone says, "I spot-checked a few and they looked fine." That sentence is the Pixie's calling card.

Emergence Conditions

High trust in AI output. Someone on the team has been using Copilot or Claude for a few weeks. Everything it's produced has been solid. They stop reading the output line by line. Why would they? It's been right every time. That trust is the Pixie's favorite entrance. The Faithful Automator pattern from the Cleric class is its preferred host: someone who believes the tools are usually right and therefore doesn't need to verify.

Absence of deterministic guardrails. When there are no unit tests, no type checking, no linting, no integration tests; when the only validation is "does it look right?", the Pixie has free rein. Its dust settles on everything because nothing brushes it off.

Work outside core expertise. The Pixie is most dangerous when the AI is working in a domain that's adjacent to the developer's expertise but not squarely within it. A backend developer generating frontend validation logic. A Python developer generating Rust memory management. The developer knows enough to read the output but not enough to catch the subtle wrongness.

Volume outpacing review. When the team generates code faster than they review it, the Pixie accumulates. Each unreviewed output is a place where the dust might have landed wrong, and at the volumes AI tools enable, 6% is a lot of bugs.

Compounding errors. The Pixie's most insidious emergence condition: the output of one AI generation becomes the input to another. Each hop compounds the probability of error. If each step is 94% accurate and you chain five steps, your end-to-end accuracy is about 73%. Chain ten and you're below 54%. The dust accumulates.

Stats and Abilities

Attribute	Value
Type	Fey (Tiny)
Severity	5
Resilience	22 (but incredibly hard to target)
Evasion	19 (Pixie Dust Evasion, hard to catch what you can't see)
Speed	60 ft. flying
Intelligence	14 (the Pixie knows exactly what it's doing)

Mostly-Correct Dust. The Pixie sprinkles dust on AI-generated code. Affected code has a 94% chance of being correct on any given execution. The remaining 6% manifests as subtle logic errors, edge case failures, or silent data corruption. Detecting the dust requires a Difficulty 16 Detection check or a comprehensive test suite, whichever is harder to obtain.

Glamour of Plausibility. All code touched by the Pixie radiates an aura of correctness. It *looks* right. Variable names are sensible. Control flow appears sound. Code reviewers must succeed on a Difficulty 14 Judgment check or approve the PR without catching the error. The difficulty increases by 2 for every hour of code review fatigue.

Vanish. When caught (bug report filed, test failure detected), the Pixie disappears and reappears in a different module. Fixing one Pixie-dusted function does not reduce the probability of finding another. The Pixie is not a single instance; it's an environmental condition.

Chain Reaction. When Pixie-dusted output feeds into another AI generation, the error probability compounds multiplicatively. The Pixie doesn't need to attack directly; it just waits for the pipeline to amplify its work.

Damage Profile

Subtle production bugs. The Pixie's bugs don't crash the system. They produce *almost* correct results. Financial calculations that are off by a fraction of a percent.

Search results that are relevant but in the wrong order. Timestamps that are correct in one timezone and wrong in another. These bugs are almost impossible to catch because the output is plausible.

Erosion of testing confidence. Once the team discovers a Pixie-dusted bug, they start questioning everything. If *that* function was wrong, what else is wrong? Trust in the codebase drops. Code review becomes paranoid and slow. The velocity gains from AI generation evaporate as the team rechecks everything.

Edge case graveyards. The Pixie consistently fails on edge cases, including nulls, empty strings, boundary values, Unicode, leap years, timezone transitions. AI training data handles these inconsistently, and developers rarely think to test them explicitly. The result: a codebase that works beautifully on the happy path and falls apart at the margins.

Compounding inaccuracy. In data pipelines, the Pixie's damage multiplies. Each transformation step introduces a small error. By the end of the pipeline, the output has drifted far enough from correctness to cause real damage, but each individual step looks fine in isolation.

Counter-Play

The Fighter's testing wall. The Bulldozer of Confidence is the Pixie's worst enemy, not in its usual mode of pushing code through, but when it insists on comprehensive test coverage as a non-negotiable. The Fighter who says "nothing ships without edge case tests, period" and refuses to budge is building a net that catches the Pixie's dust. The Daredevil of Disaster's reckless energy, turned around, is exactly the stubbornness you want in someone who refuses to ship untested code.

The Wizard's type system. The Illusionist of Complexity can be redirected toward building a type system that encodes business invariants. If the types won't let you represent invalid states, the Pixie's dust can't settle. Static analysis, strong typing, and compile-time checks are the Wizard's wards against probabilistic error. Make the incorrect states unrepresentable.

The Rogue's spot checks. The Bandit of Borrowed Brilliance, when aimed constructively, becomes a sampling auditor. Rogues are fast. They can spot-check AI output with a rapidity that slower archetypes can't match. Random sampling of AI-generated code (pulling five functions a week and verifying them by hand) creates a statistical counter to the Pixie's statistical mischief.

The Cleric's verification ritual. The Faithful Automator is the Pixie's favorite victim, but the Ritualist of Redundancy holds the cure. Establish a ritual: every AI-generated function gets a manually written test. Not an AI-generated test — a human-written test that encodes human understanding of what the function should do. This is tedious. It is also the single most effective Pixie countermeasure in this guide.

Practical counter-measures:

- Never chain AI outputs without intermediate validation. If step A feeds step B, verify A's output before B consumes it.
- Build edge case test suites by hand. These are the tests the AI won't write well because the training data underrepresents edge cases.
- Use property-based testing and fuzzing to explore the input space beyond what example-based tests cover.
- Track error rates over time. If you're catching bugs in AI-generated code, measure the rate. If it's above your tolerance threshold, slow down and increase review coverage.
- For critical code paths (financial calculations, security logic, data integrity), do not use AI-generated code without line-by-line human review. The inconvenience is the point.

Phantom Intern

The ghost of a departed developer still haunts an elegant module nobody wants to touch.

Description

You can feel the Phantom Intern's presence before you see its work. It's that moment in a code review when you open a file, read twenty lines, and think: "This is... actually brilliant. Who wrote this?" You check git blame. It was Sarah. Sarah left the company eight months ago. Sarah's commit messages are immaculate. Sarah's code is elegant, concise, and uses patterns you've never seen in this codebase.

Sarah also did about 80% of her work through AI code generation, left no documentation explaining her architectural decisions, and the Slack channel where she discussed her approach was archived when she departed.

This is where the Phantom Intern does its real damage. Complex software systems don't spring into existence fully formed. They evolve over years, shaped by decisions that anticipate features not yet built, constraints not yet encountered, and requirements that haven't been articulated yet. A developer working on a system in March might structure something a particular way because they know a major integration is coming in September. An architect might accept a subtle performance trade-off because they expect the data model to shift in a way that makes it irrelevant. The code alone doesn't tell you any of this. It tells you *what* was built, not *why* it was built that way, or what the person who built it was thinking about when they made the call.

When the Phantom Intern strikes, you end up with code that can't tell its own story. The implementation is sound. The tests pass. But the narrative thread that connects this module to the system around it — the reasoning, the context, the forward-looking decisions — is gone. It lived in prompts that were never saved, in iterative conversations with an AI that has no memory, in the contributor's head which is now solving problems at a different company.

What remains is a spectral codebase: functional, elegant, and deeply spooky. Nobody wants to touch it. Nobody fully understands it. Everyone routes around it the way you'd route around a room in the house where the lights flicker and the temperature drops. It works, so you leave it alone. But "works and nobody understands it" is just "breaks and nobody can fix it" on a delayed timer.

The Phantom Intern differs from traditional knowledge-silo problems in one specific way: the code itself is often better than what the team would produce. This creates a psychological trap. You can't dismiss it as bad code that needs rewriting. It's *good* code that needs understanding. And understanding code you didn't write, that was generated through a long conversation you'll never see, is a particular kind of archaeological challenge.

Emergence Conditions

The Phantom tends to form under a specific set of conditions. None of them are unusual on their own; most are routine in teams that adopted AI tooling quickly.

AI-heavy individual contributors. The Phantom forms when a developer relies heavily on AI tools and produces code that reflects the AI's style more than the team's conventions. During their tenure that is not necessarily a problem; the code works, after all. The trouble materializes when they leave.

Low documentation culture. If the team doesn't require design documents, architecture decision records, or meaningful commit messages, the Phantom leaves no trace until it's too late. The Rogue's Bandit of Borrowed Brilliance accelerates this: a fast-moving developer pulling brilliant solutions from AI sessions won't slow down to document unless forced.

Single-owner modules. When one developer owns a module exclusively (nobody else reviews it, nobody else contributes to it), all the understanding lives in one head. Add AI-assisted generation to the mix and the situation gets worse: even that one developer might not remember why the code works the way it does, since the reasoning lived in a prompt conversation that's been garbage-collected.

High turnover in technical roles. The Phantom Intern becomes a recurring haunting in organizations with high turnover. Each departing developer leaves their own ghost in the codebase. Eventually, you have a team of living developers maintaining code written by a team of ghosts.

No prompt or decision archiving. If the AI conversation that produced the code isn't captured anywhere (and it almost never is), then the generation context is permanently lost. The code is an artifact without provenance, a conclusion without an argument.

Stats and Abilities

Attribute	Value
Type	Undead (Incorporeal)
Severity	7
Resilience	N/A, you cannot kill what has already left
Evasion	20 (Spectral Evasion, can't hit what you can't comprehend)
Speed	0 ft. (the Phantom doesn't move; it haunts a fixed location)
Intelligence	16 (the code is genuinely clever, which is the problem)

Haunted Module. The Phantom occupies a module or service, making it resistant to modification. Any developer attempting to change the haunted code must make a Difficulty 15 Comprehension check to understand the existing logic before modifying it. On a failure, their change introduces a regression that doesn't surface for 1d4 sprints.

Spectral Brilliance. The Phantom's code radiates an aura of competence that discourages refactoring. Team members instinctively route around the haunted module rather than attempting to understand it. This aura imposes disadvantage on all "let's rewrite this" persuasion checks.

Whisper of Context. Occasionally, a developer will read a comment or variable name that hints at the original reasoning. These whispers are tantalizing but incomplete, enough to prevent a full rewrite but not enough to enable confident modification. The Phantom keeps the team in a state of permanent half-understanding.

Possession. When a new developer joins the team and tries to understand the Phantom's code, they sometimes start replicating the Phantom's AI-dependent workflow, absorbing the style without the context, and quietly setting up the conditions for a future haunting. The cycle continues.

Damage Profile

The Phantom's damage is structural and slow to show up.

Frozen architecture. The Phantom's module becomes unchangeable. The team builds around it, which distorts the architecture over time. Dependencies accumulate. Workarounds multiply. The haunted module becomes load-bearing not because of its design but because of the team's fear of touching it.

Knowledge vacuum. With the Phantom gone, nobody on the team understands their module. In a crisis (a production bug, a security vulnerability, a required API migration), there is no one who can safely modify the code. Emergency fixes become high-risk operations performed with crossed fingers and rollback scripts.

Onboarding friction. New team members hit the Phantom's code and stall. They ask questions nobody can answer. They read code that follows conventions from no known style guide. Their ramp-up time doubles, and their confidence in the codebase drops before they've written their first line.

Invisible technical debt. The Phantom's code is not technically debt in the usual sense; it works, and it is often well-written. But code that cannot be maintained is debt by another name. The interest shows up in delayed feature delivery, risky hotfixes, and the slow erosion of team confidence.

Counter-Play

Each class brings a different approach to an exorcism, and the Phantom is one of the few enemies that benefits from all four working together.

The Fighter's insistence on documentation. The Bulldozer of Confidence should be aimed at a simple policy: no module ships without a design document or architecture decision record. Not a novel. A page. Why does this module exist? What are its key decisions? What are the non-obvious trade-offs? The Fighter's force is most useful before the Phantom forms, as a prevention measure. Make documentation a non-negotiable condition of merge approval.

The Wizard's code archaeology. The Archmage of Analysis Paralysis is usually a liability, but the Phantom Intern is one of the few enemies where thorough analysis is genuinely warranted. A Wizard who systematically reads, annotates, and documents the Phantom's code (inferring the decisions from the implementation) can perform an exorcism. It's slow, careful work. It's exactly the Wizard's strength.

The Rogue's pragmatic boundary. If the Phantom's module is truly impenetrable, the Rogue's instinct to isolate and contain is correct. Wrap the haunted module in an interface. Write integration tests that treat it as a black box. Don't try to understand the interior. Just make sure you know exactly what goes in and what comes out. The Shadow Patcher of Ephemeral Solutions, for once, is the right call: patch around the ghost and move on.

The Cleric's knowledge ritual. The Doomsday Archivist pattern (obsessive documentation and archiving) is the preventive cure. Establish a ritual: when a developer leaves, their final two weeks include a knowledge transfer sprint. Every module they own gets documented. Every non-obvious decision gets recorded. This is the Cleric's ceremony aimed at the right problem: capturing context before it evaporates.

A few structural practices reduce the Phantom's chances of forming, or limit the damage when it does anyway:

- Archive AI conversations alongside code. If the reasoning lived in a prompt session, save that session as part of the project record.
- Require pair programming or cross-review on AI-heavy modules. The Phantom forms when only one person understands the code. Make sure at least two people do.
- Run regular "code understanding" sessions where developers present modules they didn't write. If nobody can explain a module, flag it for documentation before the author leaves.
- When a team member gives notice, treat their codebase knowledge as a perishable asset. Schedule extraction sessions immediately, not during their last week.

The Phantom Intern is never malicious. They were productive, collaborative, and genuinely good at what they did. The haunting is just what happens when context walks out the door and nobody thought to write it down first.

Stagnant Miasma

A comfortable fog of repeated patterns settles over a team that stopped questioning.

Description

The Stagnant Miasma doesn't attack. It settles.

It's a faint, comforting haze that smells like familiarity: the patterns you already know, the tools you've already mastered, the architectures you've already shipped. It whispers that what worked last time will work this time. And if the AI tools you're using happen to reinforce those existing patterns, well, that just confirms you were right all along.

The Miasma is stagnation dressed as stability. Large language models are trained on historical data. They are, by their nature, backward-looking. When you ask an AI to generate code, it draws from patterns that exist in its training corpus, which means it reflects the collective conventions, architectures, and practices of the past. This is usually fine. Most software doesn't need to be novel. But when a team uses AI tools exclusively to replicate what they've already done, something insidious happens. The AI validates their existing approach, the team's confidence in their current patterns increases, and new ideas feel riskier because the AI doesn't suggest them. Innovation slows. Then it stops.

The Miasma is particularly dangerous because it feels like wisdom. Experienced engineers know that chasing novelty for its own sake is foolish. "Don't fix what isn't broken" is genuinely good advice most of the time. The Miasma exploits this legitimate instinct and stretches it past the breaking point, into territory where "don't fix what isn't broken" becomes "don't change anything, ever, because change is risk and we have a tool that makes the old way even easier."

There's a mechanism that accelerates the Miasma in ways most teams don't realize. Modern AI coding tools don't just respond to your prompts. They respond to the context you've built around them. Claude Skills, Cursor Commands, AGENT.md files, system prompts, project instructions: these are all ways of telling the model how to behave. And many teams, with the best of intentions, load these files with restrictive instructions. "Always follow our existing patterns." "Use our established conventions." "Never suggest architectural changes without approval." The model does exactly what it's told. It generates code that fits neatly within the boundaries the team defined, and the team interprets this as confirmation that their boundaries are correct. But the model isn't agreeing with the architecture. It's complying with the instructions. The team has built a cage and then congratulated themselves on how well the bird stays inside.

There's a subtler trap that comes from misunderstanding what these models actually are. Teams personify them. They talk about "the AI" the way they'd talk about a senior engineer: "the AI didn't flag any issues with this approach," or "the model seems comfortable with our architecture." This language implies judgment, evaluation, independent thought. It implies the model looked at the situation and formed an opinion. That's not what happened. The model identified patterns in the input and predicted the most likely next tokens. It has no opinion. It has no authority. It has no awareness that your architecture exists outside the context window you just handed it.

This matters because of what happens when the model's constrained output gets cited as evidence. A team deep in the Miasma will point to the model's silence as proof of correctness. "Well, the model didn't suggest any changes to this module." Maybe. But did anyone check whether the AGENT.md file sitting in the repo root has a directive telling the model not to suggest changes? Did anyone notice that the system prompt loaded into their IDE says "always follow existing patterns"? The model didn't evaluate the architecture and find it sound. The model was told not to evaluate the architecture, and it complied. The team then cites that compliance as independent validation. They've built a confirmation machine and mistaken it for an advisor.

The danger isn't that the model is wrong. The danger is that the team has outsourced their own judgment to a system that is, by design, incapable of exercising it. The model cannot tell you that your architecture needs to change if your instructions tell it not to suggest changes. It cannot flag a growing problem if your context files frame the current approach as proven and mature. The model is not your senior engineer. It's a very capable pattern-matcher operating inside the fences you built. If you never check whether the fences are still in the right place, the model certainly won't.

Then there's the sycophancy problem. Large language models are trained, in part, to be helpful and agreeable. When a team feeds the model context that frames their current approach in positive terms, the model tends to reflect that framing back at them. Tell a model that your microservices architecture is battle-tested and the model will generate code that assumes microservices are the right call. Describe your team's patterns as "mature and proven" and the model will treat them as constraints to work within, not assumptions to question. The model isn't being dishonest. It's doing what it was designed to do: produce output that aligns with the context it receives. But the effect is a feedback loop. The team tells the model their approach is great. The model generates code consistent with a great approach. The team reads the output and thinks, "See? Even the AI thinks we're on the right track." Nobody in this loop is lying. Everyone is wrong.

Emergence Conditions

Successful prior projects. The Miasma doesn't form in teams that are failing. It forms in teams that have *succeeded*, and concluded that their success means their approach is permanently correct. Success calcifies. The Shield Bearer of Stagnation from the Fighter class is the Miasma's ideal carrier: someone who defended the current approach so effectively that alternatives stopped being considered.

AI tools that reflect back existing patterns. When a team's prompts consistently produce code that looks like their existing codebase, they interpret this as validation. "Even the AI agrees with our architecture." What's actually happening is that the AI is pattern-matching to their prompts, which are framed in their existing vocabulary, which produces output that resembles their existing code. It's an echo chamber with a neural network in the middle.

Restrictive context files. Claude Skills, Cursor Commands, AGENT.md, and similar configuration files are powerful tools for steering AI behavior. They're also the Miasma's favorite habitat. When teams load these files with instructions that constrain the model to existing patterns, forbid architectural suggestions, or require strict adherence to current conventions, they've effectively told the AI to never challenge the status quo. The model complies. The team never sees an alternative. The fog thickens.

Sycophantic feedback loops. AI models are trained to be helpful and agreeable. When a team's context and prompts frame their current approach in positive terms, the model mirrors that confidence back at them. The team reads the model's output as independent validation when it's actually just reflection. Over time, the team develops a dependency on AI-generated confirmation that their choices are correct, and the model keeps providing it because the team keeps asking for it.

Risk-averse organizational culture. The Miasma thickens in environments where failure is punished more than stagnation. If trying something new and failing has consequences but maintaining the status quo doesn't, engineers choose the status quo. AI tools make this even easier. You can be productive and stagnant simultaneously.

Long-tenured, homogeneous teams. Teams where everyone has been around for years and shares the same technical background are the Miasma's natural habitat. There's no outsider perspective to challenge the fog. New hires absorb the existing patterns through osmosis and AI-assisted onboarding that teaches them the current way, not a better way.

Absence of external pressure. When there's no competitive threat, no evolving customer demand, no technological shift demanding adaptation, the Miasma settles comfortably. The urgency to innovate evaporates, and the fog fills the vacuum.

Stats and Abilities

Attribute	Value
Type	Elemental (Air/Poison)
Severity	6
Resilience	95 (resilient but dispersible)
Evasion	10 (you can't miss it, but you can't sword-fight fog)
Speed	0 ft. (it doesn't move; it permeates)
Intelligence	8 (just smart enough to whisper plausible justifications)

Soporific Cloud. All creatures within the Miasma must make a Difficulty 13 Judgment check at the start of each sprint or become *complacent*. Complacent creatures automatically reject proposals for new tools, patterns, or approaches. They gain advantage on all "we should keep doing what we're doing" persuasion checks.

Pattern Reinforcement. When an AI tool generates code within the Miasma, it produces output that matches the team's existing patterns 100% of the time, regardless of whether those patterns are appropriate for the current problem. The Miasma doesn't make the AI worse; it makes the team's prompts narrower.

Fog of Competence. The Miasma creates a false sense of expertise. Team members feel productive. Metrics confirm their feelings. The fog obscures the distinction between "we're efficient at what we do" and "what we do is the right thing to be doing." These are different claims. The Miasma conflates them.

Resistance to Ventilation. When someone attempts to introduce a new approach, the Miasma thickens around the team. Arguments against change become more articulate. Historical examples of failed innovation are summoned with suspicious clarity. The Miasma is self-defending.

Damage Profile

Competitive obsolescence. While the team iterates on their proven patterns, the industry moves on. New tools, new paradigms, new capabilities pass them by. The damage isn't visible until a competitor ships something the team couldn't build because they never learned the techniques.

Talent drain. Ambitious engineers leave. They can feel the Miasma even if they can't name it. The first sign is senior hires who depart within a year, citing "lack of growth opportunities." The second sign is that the only people who stay are the ones who find the fog comfortable.

Architectural brittleness. Systems built on unchanging patterns accumulate hidden fragility. The patterns were designed for a context that no longer exists. Requirements have shifted, the tools and paradigms have moved on. But the architecture hasn't, because the Miasma made adaptation feel unnecessary. When the breaking point comes (and it will), the system shatters along fault lines nobody noticed because the fog obscured them.

Innovation debt. The Miasma's most expensive damage is the compound interest on unexplored alternatives. Every quarter that passes without evaluating new approaches is a quarter of learning that did not happen. Unlike technical debt, innovation debt does not show up in the codebase. It shows up in the team's diminishing ability to respond to change.

Counter-Play

The Fighter's rebellion. The Shield Bearer of Stagnation is the Miasma's avatar within the team. To counter it, the Fighter must reverse polarity, channeling the Daredevil of Disaster's willingness to take risks, but in a controlled, deliberate way. Schedule "demolition days" where the team intentionally builds something using tools and patterns they've never tried. The Fighter's force cuts through fog when aimed at inertia instead of defending it.

The Wizard's exploration mandate. The Visionary of Grandiose Plans is a liability in most contexts, but against the Miasma, the Wizard's tendency to think beyond the current horizon is medicinal. Give the Wizard explicit license to research and prototype with new approaches. Time-box it. Don't let it become Analysis Paralysis. But let the Wizard bring back one new technique per quarter and present it honestly, covering strengths, weaknesses, and a credible comparison to the current approach.

The Rogue's external intelligence. The Bandit of Borrowed Brilliance, reframed constructively, is the Miasma's natural predator. Rogues read widely, attend conferences, and lurk in forums and open-source communities. They know what other teams are doing. A Rogue who brings external patterns into the team's awareness (not as mandates but as options) ventilates the fog. "Here's how Company X solved this differently" is a powerful antidote to "this is how we do things."

The Cleric's retrospective. The Ritualist of Redundancy can be weaponized against the Miasma through honest retrospectives. Not the ceremonial kind where everyone says things are fine, the uncomfortable kind where someone asks "what have we learned this quarter that we didn't know last quarter?" If the answer is "nothing," the Miasma is in the room. The Cleric's process discipline should include a standing ritual of discomfort: regularly questioning whether the current approach is still the right one.

Practical counter-measures:

- Institute a "technology radar" review every quarter. What's new? What's changed? What should we evaluate? The review doesn't need to produce action items; it needs to produce awareness.
- Hire for cognitive diversity. Bring in developers with different backgrounds, different tool preferences, different architectural instincts. The Miasma can't survive cross-ventilation.
- Rotate AI prompting styles. If every developer prompts in the same way, using the same vocabulary, the AI will keep reflecting the same patterns. Experiment with different framings. Ask the AI for alternative approaches explicitly.

- Audit your context files regularly. Claude Skills, Cursor Commands, AGENT.md, and similar configuration files should be reviewed for restrictive language that tells the model to stay inside existing patterns. If your instructions say "always follow our conventions," ask yourself whether those conventions are still the right ones. The model can't challenge assumptions you've told it to accept.
- Periodically ask the model to critique your architecture, not just extend it. Frame the prompt as a deliberate exercise: "What would you change about this codebase if you had no constraints?" The answers won't all be good, but they'll break the feedback loop.
- Track the age of your core patterns. If your fundamental architecture hasn't changed in three years, that's not necessarily a problem, but it *is* a question worth asking.

Temporal Tangle Worm

A giant worm threads through roadmap time, trapping the present between sunk costs and imagined futures.

Description

The Temporal Tangle Worm doesn't live in your codebase. It lives in your roadmap.

It slithers through time, coiled around past decisions you cannot easily undo and future promises you have not earned yet. Some of what feeds it starts as good judgment: teams should think ahead, notice when the tools are changing quickly, and avoid painting themselves into corners. The Worm appears when sensible foresight turns into devotion to a future version of the product, the platform, or the model stack that does not exist yet.

The future is arriving faster with generative AI than it has with any previous wave of technology. This is not a metaphor. The tools themselves are accelerating software delivery, which means the competitive landscape reshuffles every few months instead of every few years. A team that spent six weeks building a product around a specific model capability can wake up to find that a new release has made their product redundant. A team that carefully priced out the cost of leveraging generative AI to enable a feature in January can discover by February that an entirely new set of models, tools, and pricing tiers has rendered their cost model obsolete. The ground doesn't just shift under you. It shifts while you're standing on it, and the Worm feeds on the vertigo.

This is the siren song of future innovation, and it is singing to every team working with generative AI right now. Should we wait for the next model release before committing to an architecture? What's the right way to build against agentic security standards when those standards are still being written? What's happening to MCP, and should we bet our integration strategy on it or wait to see if it stabilizes? Should we hold the release another month to see what OpenAI announces next? These are not hypothetical questions. They are the questions being asked in planning meetings across the industry, and every one of them is a segment of the Worm's body.

The Temporal Tangle Worm captures roughly half of the projects currently developing with generative AI. That is not an exaggeration. Walk into any planning meeting for an AI-augmented product and you will find teams paralyzed by the same set of temporal traps: the sunk cost of decisions made against a model that has since been superseded, the fear of committing to an approach that might be obsolete in weeks, the temptation to delay shipping in hopes that the next release will make the hard problems disappear. The Worm thrives in this environment because the environment itself is genuinely uncertain. The capabilities are real. The pace of change is real. The paralysis is real. And the cost of waiting is just as real as the cost of committing to the wrong thing.

The Worm's core trick is making tomorrow's possibilities feel like today's obligations. "We should build for GPT-11 compatibility" becomes a real line item even though GPT-11 does not exist. "When the model gets better at code review, we will automate the whole pipeline" becomes a reason not to fix the pipeline the team already has. The roadmap starts drifting toward a future state that keeps receding because AI capabilities move faster than anyone can plan with much confidence.

Emergence Conditions

The Worm tends to appear when a few understandable team habits start reinforcing one another:

Rapid AI capability evolution. The Worm feeds on change. When new model releases land every few months, each one with genuinely impressive new capabilities, teams start planning around future releases instead of current ones. "We'll wait for the next version" becomes a permanent posture.

Grandiose vision without immediate accountability. Long-range thinking matters. The Worm thrives when leadership rewards the long-range part but does not check what is actually shipping in the near term. If the team can earn praise for a brilliant roadmap while shipping nothing this quarter, the Worm grows fat and comfortable.

Technical debt used as an excuse for postponement. "We need to rebuild the foundation before we can add features" is sometimes true and sometimes the Worm speaking. When the rebuild timeline keeps stretching because the target architecture keeps changing to accommodate hypothetical AI capabilities, the team is already tangled.

Sunk cost entanglement. Past decisions (the framework you chose, the model you integrated, the architecture you committed to) become segments of the Worm that restrict present movement. "We already invested in this approach" makes it hard to change course, even when the approach is no longer optimal. The past segments of the Worm are just as binding as the future ones.

Demo-driven development. When the primary output of a sprint is a demo of what *could* be built rather than a release of what *has* been built, the Worm has established residence. Demos show possibility. Shipping shows capability. The Worm prefers possibility because it is shinier and asks for less commitment.

Stats and Abilities

The Worm has one reliable move: it turns uncertainty about the future into permission to neglect the present.

Attribute	Value
Type	Aberration (Temporal)
Severity	8
Resilience	130 (one segment for each deferred decision)
Evasion	15 (hard to attack something that technically doesn't exist yet)
Speed	Variable (moves freely through past, present, and future)
Intelligence	12 (cunning enough to show you exactly the future you want to see)

Temporal Displacement. At the start of each sprint, the Worm forces a Difficulty 14 Judgment check. On a failure, the team spends 25% of their capacity on future-state planning instead of current-state building. This compounds: two consecutive failures mean 50% of capacity is tangled in tomorrow.

Shimmer Lure. The Worm projects visions of future AI capabilities that are just plausible enough to be irresistible. "Imagine when models can self-test." Everyone imagines it. Nobody ships the manual test suite that's needed right now. The Shimmer Lure imposes disadvantage on all "let's just build the boring thing that works today" arguments.

Retroactive Binding. Past decisions become load-bearing segments of the Worm. Each sunk cost increases the difficulty of "let's change direction" checks by 1. A project with five quarters of investment requires a Difficulty 19 Judgment check to change course, even if changing course is obviously correct.

Paradox Coil. When the team tries to plan for both present needs and future capabilities simultaneously, the Worm creates a paradox: the present architecture is too simple for the future vision, and the future architecture is too complex for present needs. Neither gets built. The team oscillates between scopes, making progress on neither.

Damage Profile

The Worm does not usually create dramatic failure all at once. It creates a long season of neglect, the kind everyone can explain and nobody can justify.

Present-tense neglect. The most direct damage: features that users need now don't get built because the team is architecting for a future nobody can predict. Customer complaints accumulate. Churn increases. The team is building for a future that requires the customers they're currently losing.

Perpetual beta. Nothing feels done. Every release is a stepping stone to the *real* release. V1 is just a foundation for V2, V2 is just preparation for the AI-native V3. The product exists in a permanent state of becoming, never arriving.

Decision paralysis. Should the team use the current model API or wait for the next version? Should they build the feature manually or wait for the AI to get better at generating it? Every decision picks up a time dimension that makes commitment feel premature. The Worm creates an environment where "let's wait and see" is always the easiest answer.

Roadmap fiction. The roadmap starts changing with every model announcement, every capability preview, every blog post from an AI lab. It stops being a plan and starts behaving like a mood board. Teams that have been tangled for long enough stop distinguishing between "we are planning to do this" and "we hope we will be able to do this if the technology cooperates."

Team demoralization. Engineers want to ship. When the goalposts move every quarter because the future state keeps shifting, motivation erodes. The best developers start looking for teams that actually deploy code to production rather than deploying slide decks about code that will be deployed to production eventually.

Counter-Play

The Fighter's present-tense commitment. The Daredevil of Disaster is useful here, not for recklessness, but for the willingness to commit to *now*. The Fighter who says "we ship this Friday with this model and this architecture, full stop" cuts through the Worm's temporal displacement. The Bulldozer of Confidence, aimed at a near-term deadline rather than a grand vision, is the Worm's primary antagonist.

Deadlines are temporal anchors. The Worm can't displace a team that's locked to a date.

The Wizard's scope reduction. The Archmage of Analysis Paralysis is the Worm's best friend, unless the Wizard turns that analytical energy toward ruthless scope reduction. A Wizard who says "given the tools we have *today*, what can we ship in two weeks?" and refuses to entertain hypothetical future capabilities is doing the team a real service. Constrain the analysis to present-tense inputs and the Wizard produces present-tense plans.

The Rogue's prototype discipline. The Shadow Patcher of Ephemeral Solutions is the Worm's antivenom when applied correctly. Instead of planning for the future, the Rogue builds a quick prototype with today's tools. It is usually a little ugly, but it works and users can touch it. Now the team has a real thing instead of a shimmering vision, and real things generate feedback that anchors the roadmap in reality.

The Cleric's sprint boundary. The Ritualist of Redundancy, when aimed at sprint commitments rather than arbitrary process, creates a temporal ward. The ritual is simple: at the start of each sprint, define what ships at the end. No future-state features. No "laying groundwork for next quarter." Only work that produces user-facing value within the sprint boundary. The Cleric enforces the boundary with the solemnity of a religious vow, and that is not a joke about process. That is the actual level of commitment required.

The practical counter-moves are not glamorous:

- Ban the phrase "when the next model comes out" from planning meetings. Plan with the tools you have, not the tools you hope for.
- Separate exploration from execution. Give the team a fixed, small budget for investigating future capabilities. Everything else ships against current requirements.
- Track the ratio of shipped features to planned features. If the number is shrinking, the Worm is growing.
- Set immovable ship dates. Not arbitrary deadlines. Genuine commitments to real users. The Worm can't thrive in the presence of a customer waiting for a delivery.

- When evaluating an architectural decision, ask: "does this solve a problem we have today, or a problem we think we'll have?" If the answer is the latter, table it.

How Monsters Combine: Multi-Failure Scenarios

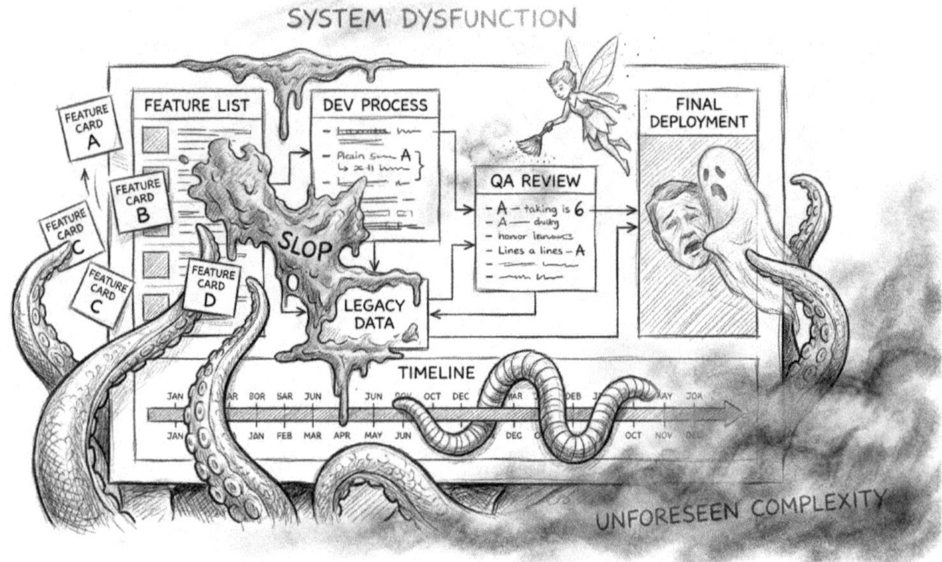

Every monster in the field guide converges on one architecture diagram at once.

Monsters in the field guide do not respect each other's territory. They overlap, compound, and amplify each other in ways that make the individual entries look almost quaint. A Scope Creep Kraken is bad. A Scope Creep Kraken swimming through Congealing Slop is how a small project turns into a recovery effort.

This chapter sketches three multi-failure scenarios drawn from real team dynamics. They are composites, not case studies, but every engineer who has lived through one will recognize the shape.

The Drowned Architecture: Scope Creep Kraken + Congealing Slop

The team started with a clean brief: build an internal tool for the data engineering team to validate pipeline configurations. Two engineers, six weeks, straightforward scope. They chose an AI coding assistant to accelerate development, and the first two weeks were spectacular. The core validation logic was done. The CLI worked. The tests passed. The project was ahead of schedule.

Then someone ran a demo for the broader engineering org.

"Could it also validate our Kafka topic configurations?" Sure. The AI generated a Kafka validator in twenty minutes. "What about our Terraform modules?" Another afternoon. "Could we add a web UI instead of just the CLI?" The AI scaffolded a React frontend by end of day. The Scope Creep Kraken had arrived, and the team was thrilled to feed it. Each tentacle was genuinely useful. Each addition solved a real problem. The two-person project became a platform initiative with seven feature tracks and no additional headcount.

Each new feature was generated in its own AI session, with its own prompting context, by whichever developer happened to pick up the task. The Kafka validator used a different error-handling pattern than the Terraform validator. The web UI made assumptions about the API response format that didn't match what the CLI expected. The original validation logic used synchronous calls; the newer modules were async. Nobody noticed because each piece worked in isolation.

By week eight, the team had a platform that could do seven things, none of which worked reliably when combined. The web UI showed stale results because the async validators resolved in a different order than the synchronous originals. The Terraform module crashed when it encountered a Kafka configuration block that the Kafka validator handled fine. The test suite passed, because each module had its own tests, written by the same AI that wrote the code, validating the same assumptions.

What makes this combination dangerous: The Kraken provides the surface area. The Slop fills it with incoherent code. Without the Kraken, the Slop would be contained to a small codebase where inconsistencies are manageable. Without the Slop, the Kraken's tentacles would at least be internally consistent. Together, you get a large, sprawling system where every module disagrees with every other module about fundamental conventions. Refactoring is impossible because the scope is too large. Scoping down is impossible because the modules are too entangled.

Counter-play focus: You must fight the Kraken first. Cut scope ruthlessly. Identify the two or three features that serve the original mission and deprecate the rest.

Only after the scope is contained can you address the Slop, because cleaning up code in a codebase that's still expanding is like mopping during a flood.

The Fighter holds the scope line, picks the three features that matter and says no to everything else. The Wizard audits integration points between modules, looking for the places where conventions silently diverge. Somewhere in the branch history, there are merges that never should have happened; the Rogue finds them and reverts. And the Cleric institutes a freeze: nothing new ships until what exists is coherent.

The Confident Ghost: Probability Pixie + Phantom Intern

Marcus was the staff engineer who joined the team, spent four months building the ML inference pipeline, and left for a startup. During his tenure, he used AI tools extensively, not carelessly, but confidently. His code was clean, well-structured, and accomplished complex things with elegant simplicity. His commit messages were terse. His documentation was minimal. His Slack messages about design decisions were in a DM channel with a colleague who also left. I have worked with at least two people like Marcus. You probably have too.

Six months after Marcus's departure, the pipeline started producing subtly wrong results. Not crashes. The Probability Pixie doesn't crash things. Customer-facing predictions were off by small margins. Confidence scores didn't match actual accuracy. The drift was gradual enough that it took three weeks to confirm it wasn't just statistical noise.

The team opened Marcus's code and hit the Phantom Intern's spectral wall. The pipeline worked, but nobody could explain *why* specific preprocessing steps were ordered the way they were. There was a normalization function that seemed to apply corrections based on a statistical model that wasn't documented anywhere. Comments were sparse. The function names were descriptive but the *reasoning*, why this normalization instead of that one, why applied at this stage instead of earlier, was gone. It had lived in Marcus's conversations with Claude, iterated over dozens of prompts, refined through trial and error that was never recorded.

Now the team had a Pixie problem inside a Phantom problem. The Pixie's dust (the subtle inaccuracy) was somewhere in the pipeline, but finding it required understanding the pipeline's logic, and understanding the logic required expertise that had walked out the door. Every attempted fix was a guess. They'd adjust a threshold, and accuracy would improve for one customer segment and degrade for another. They'd modify the normalization step, and the confidence scores would shift in ways they couldn't predict because they didn't understand what the normalization was originally compensating for.

What makes this combination dangerous: A Pixie problem is fixable. You find the inaccuracy, you test it, you correct it. A Phantom problem is manageable; you can't modify the code easily, but at least it works. Put the Pixie inside the Phantom's code and the situation changes: you have inaccuracy inside a black box. You can't fix what you can't understand, and you can't understand what a ghost left behind. The debugging cycle becomes archaeological. You're not reasoning about the code, you're excavating it.

Counter-play focus: Resist the urge to fix the inference logic directly. Instead, build a test harness around the pipeline, input/output pairs that define correct behavior regardless of internal implementation. Use property-based testing to explore the input space. Once you have an external specification of what the pipeline *should* do, you can either fix Marcus's code with confidence or rewrite it entirely with a specification to test against.

The Wizard builds the harness. The Rogue writes a parallel implementation as a cross-check, a second opinion that doesn't need to be pretty, just independently correct. The Cleric establishes a monitoring ritual for drift detection, because the Pixie should never be able to hide for three weeks again. And the Fighter draws the line that matters: no one deploys inference changes without the full test suite passing, no exceptions, no "it's probably fine."

The Comfortable Paralysis: Stagnant Miasma + Temporal Tangle Worm

Nobody yells during this one. Nobody notices for months.

The team has been building enterprise integration middleware for four years. They're good at it. Their patterns are proven. Their CI/CD pipeline is mature. Their customers are satisfied, if not excited. The Stagnant Miasma settled in around year two, when the team stopped evaluating new approaches because their existing approach kept working. AI tools reinforced this: every time they prompted for code generation, the AI produced code that looked exactly like their existing codebase, because their prompts were framed in their existing vocabulary. The echo chamber was seamless.

Then the Temporal Tangle Worm arrived via a board presentation. The CTO saw a demo of an AI agent that could automatically generate integration adapters from API documentation. She came back with a vision: the team would build an AI-native integration platform that could auto-generate connectors for any API. The roadmap was rewritten. The team was excited. Finally, something new.

Except the team couldn't build it. The Miasma had atrophied their ability to think outside their established patterns. Their architecture was designed for manually-coded integrations. Their testing framework assumed human-written adapters. Their deployment pipeline couldn't handle dynamically generated code. Everything about their infrastructure was optimized for the way they'd always done things.

So they started planning the infrastructure rebuild. Which required evaluating new frameworks. Which required prototyping. Which kept getting derailed because the next model release promised better code generation capabilities, so maybe they should wait. The Temporal Tangle Worm had them: stuck between a past they couldn't leave (the Miasma's grip) and a future they couldn't reach (the Worm's shimmer). They spent two quarters producing architectural documents for a platform that required capabilities their team hadn't developed and AI models that hadn't been released.

Meanwhile, their existing customers needed bug fixes and feature requests for the current platform. These requests were deprioritized because the team was "in transition." Customer satisfaction dropped. Two key clients started evaluating competitors. The team was too busy planning the future to maintain the present, and too anchored to the past to actually build the future.

What makes this combination dangerous: A Miasma team ships but doesn't innovate. A Worm team innovates but doesn't ship. Put both monsters in the same room and you get a team stuck in the past while gazing at the future, executing in neither time frame. The Miasma has atrophied the muscles they'd need to build new things. The Worm has convinced them that building old things is beneath them. What's left is a team that's busy, stressed, and producing nothing.

Counter-play focus: Break the temporal deadlock by forcing one concrete action in the present tense, not a plan, not a prototype for the future platform, but a single shippable feature for the current platform that uses one new technique. A small crack in the Miasma. A small anchor against the Worm.

The Fighter picks a deadline and defends it against scope and abstraction alike. The Wizard evaluates exactly one new technology (not five, not a matrix, one) and recommends adopt or reject within two weeks. The Rogue builds a proof-of-concept so ugly that nobody can mistake it for a roadmap and so functional that nobody can dismiss it as a toy. And the Cleric reinstates the sprint boundary: two weeks, shippable increment, no future-state work.

If the team can't ship something in two weeks, the problem isn't the technology. It's the monsters.

The field guide catalogs monsters individually because that's how they're easiest to recognize. But they rarely arrive alone. In practice, they layer, reinforce each other, and exploit the gaps that a single monster leaves open. A team that spots a Kraken should check whether the Slop is already congealing in its wake. A team that exorcises a Phantom should ask whether the Pixie was living inside it.

Monsters combine because the conditions that summon one usually summon others. The counter-play, when it works, follows the same principle: no single class handles a multi-failure scenario alone. The party matters, and AI tools just made the encounters faster, which means the party needs to be paying attention sooner.

Part IV: Practice

This part turns the framework back toward real team behavior. It focuses on diagnosis, intervention, and the habits that keep slop from hardening into culture once you know what to look for.

Chapter 15: Diagnosis and Intervention — A diagnostic field reference for spotting slop patterns early, paired with concrete team practices, review habits, and prompt discipline that reduce recurring failure modes.

Chapter 16: Conclusion: Build Better Habits, Not Better Excuses — Closes by reframing the goal: not perfect tools or perfect people, but stronger habits for working with AI responsibly.

By the end of this part, you will have a practical playbook for diagnosing issues early and responding with habits that improve team judgment over time.

Diagnosis and Intervention

A table of countermeasures gathers the tools needed to fight back.

Naming dysfunction is useful, but it is not enough on its own. You also need to spot the symptoms early, and you need counter-moves.

The first half of this chapter is a diagnostic field reference: **Observable Symptom -> Likely Pattern**. When something feels off, when the code reviews are too quiet, when the velocity numbers do not match the quality you are seeing, when your gut says "this isn't right" but you cannot name why, find the symptom, follow it to the pattern.

The second half gives you interventions: reviewing practices, prompt discipline, validation techniques, and retrospective questions you can put to work immediately.

Diagnostic Field Reference

Fighter Class Symptoms

Observable Symptom	Likely Pattern
PRs ship with minimal or no review; author merges their own code	Bulldozer of Confidence
Debugging sessions consist of re-generating code until tests pass	Bulldozer of Confidence
AI suggestions accepted wholesale with no modification	Bulldozer of Confidence
Team refuses to adopt new AI tools despite demonstrated value	Shield Bearer of Stagnation
Code reviews reject AI-generated code on principle, not substance	Shield Bearer of Stagnation
Production deployments happen without staging or canary testing	Daredevil of Disaster
AI-generated infrastructure code is deployed without security review	Daredevil of Disaster

Wizard Class Symptoms

Observable Symptom	Likely Pattern
Architecture discussions consume entire sprints with no code written	Archmage of Analysis Paralysis
Team requests "just one more" AI-generated comparison before deciding	Archmage of Analysis Paralysis
Nobody on the team can explain the architecture without consulting docs	Illusionist of Complexity
AI-generated code introduces design patterns that solve no current problem	Illusionist of Complexity
Roadmap includes features that no customer has requested	Visionary of Grandiose Plans

Observable Symptom	Likely Pattern
Team is building "platform" capabilities before the first product ships	Visionary of Grandiose Plans

Rogue Class Symptoms

Observable Symptom	Likely Pattern
Bugs reappear in areas that were "fixed" last sprint	Shadow Patcher of Ephemeral Solutions
AI-generated patches address symptoms without touching root causes	Shadow Patcher of Ephemeral Solutions
Code contains patterns or idioms nobody on the team recognizes	Bandit of Borrowed Brilliance
Developer can't explain why a particular approach was chosen	Bandit of Borrowed Brilliance
License compliance audits reveal unexpected copyleft dependencies	Outlaw of AI Plunder
"The AI wrote it, not me" is used to deflect attribution questions	Outlaw of AI Plunder

Cleric Class Symptoms

Observable Symptom	Likely Pattern
AI output is accepted without validation because "it's usually right"	Faithful Automator
Team trusts AI-generated tests to validate AI-generated code	Faithful Automator
Daily standups follow the same script regardless of project phase	Ritualist of Redundancy

Observable Symptom	Likely Pattern
AI tools are integrated into every workflow step whether needed or not	Ritualist of Redundancy
Storage costs are doubling quarterly with no corresponding data strategy	Doomsday Archivist
Every AI conversation, prompt, and output is logged "just in case"	Doomsday Archivist

Monster Symptoms

Observable Symptom	Likely Monster
Feature scope expands every time someone asks the AI "what else could we add?"	Scope Creep Kraken
Sprint goals shift mid-sprint because AI generated "better" ideas	Scope Creep Kraken
Codebase has inconsistent styles, conflicting patterns, and no clear conventions	Congealing Slop
Technical debt is accumulating faster than the team can name it	Congealing Slop
"The model says it should work" replaces actual testing	Probability Pixie
Risk assessments are AI-generated and never validated against historical data	Probability Pixie
Code exists that nobody on the team wrote or understands	Phantom Intern
Bus factor for critical systems is effectively zero	Phantom Intern
Innovation has stalled; the team ships the same patterns on repeat	Stagnant Miasma

Observable Symptom	Likely Monster
AI tools are used exclusively to replicate existing approaches	Stagnant Miasma
Planning meetings keep orbiting future model releases instead of current delivery	Temporal Tangle Worm
Ship dates slide because the team is "waiting for the next model"	Temporal Tangle Worm

Pick the two or three symptoms that resonated most, the ones that made you think of a specific meeting, a specific PR, a specific conversation. Read the chapters where those patterns are described in full. Then come back here for the matching interventions.

Interventions

What follows are practices I have seen work on real teams shipping real software, including teams I was on when the anti-pattern was mine. Not every intervention fits every team. Pick the ones that match your symptoms, adapt them to your context, and discard the rest. The worst thing you can do with a checklist is follow it religiously without understanding why each item exists.

Reviewing AI-Generated Code

Standard code review practices were designed for human-written code. AI-generated code requires additional scrutiny because it fails differently. Syntactically clean, idiomatically plausible, and structurally hollow in ways that only show up later.

Now for the honest part. If it is possible to review each line of code an AI generates, I am an advocate of that approach. But on the projects I have worked on, it is often unrealistic to expect a developer to review ten thousand lines of generated code in a single day and actually identify the problems. With the tools available today, no developer running four or five agents with reasoning models can honestly tell you they have reviewed every line. The volume outpaces the reviewer.

That does not mean review is obsolete. It means the nature of review has to change. What remains essential is staying connected to the design choices, the architectural approaches, and the patterns being used in the code you ship. You need to be able to explain why the system is built the way it is, not because you wrote every line, but because you understand the decisions that produced it. You don't need to review every control statement, but you do need to review the overall approach. Skipping both is not responsible — it's abdication.

"Vibe coding," the idea that you do not need to look at your code anymore, might be fine for your CTO's weekend side project. The people promoting that approach tend to be the ones who are not shipping production software for a living. If you have any sense of professional responsibility, you need to remain connected to the code being generated on your behalf and your team's behalf. These tools still need a responsible engineer behind them. That has not changed, and after the first few years of generative AI in production, it is clear it is not going to.

The "Explain It" Rule

Before any AI-generated code is approved, the submitting developer must be able to explain, in their own words, what the code does and why it was written this way. Not what the AI said it does. What the developer understands it to do.

This single practice counters three patterns at once: - **Bulldozer of Confidence** forces the developer to slow down and engage with the code before merging - **Bandit of Borrowed Brilliance** exposes cases where the developer doesn't understand what they've submitted - **Faithful Automator** breaks the habit of treating AI output as pre-validated

If the developer can't explain it, the PR goes back. No exceptions, no "I'll look at it later." The explanation happens before the merge.

AI-Specific Review Checklist

Add these to your existing code review process. They don't replace your current checklist; they supplement it.

- ☐ **Provenance declared**: The PR indicates which portions were AI-generated, AI-assisted, or human-written

- ☐ **No phantom dependencies**: All imported libraries are in the project's approved dependency list or have been evaluated
- ☐ **No speculative abstractions**: Every interface, abstract class, or pattern serves a current, concrete requirement, not a hypothetical future one
- ☐ **Tests are independent**: Tests were written or reviewed by a human, not generated by the same AI session that produced the code under test
- ☐ **Error handling is substantive**: Error paths do something meaningful, not just log and swallow or rethrow with a generic message
- ☐ **No license contamination**: Generated code doesn't introduce copyleft or incompatible license obligations
- ☐ **Naming makes sense in context**: Variable and function names reflect the project's conventions, not the AI's generic preferences
- ☐ **Complexity is justified**: If the code is more complex than the obvious solution, there's a documented reason

Counters: Bulldozer of Confidence, Illusionist of Complexity, Bandit of Borrowed Brilliance, Outlaw of AI Plunder, Phantom Intern

Prompt Discipline Practices

Most developers interact with AI tools the way they interact with search engines, fire off a query, grab the first result, move on. That habit is how Shadow Patchers form. A better model won't fix it. A better habit will: be intentional about what you ask, how you ask it, and what you do with the answer.

The Context-First Prompt

Before asking for code, provide context. Tell the model about your existing architecture, your constraints, your conventions. The difference between a generic prompt and a useful one is specificity. Compare "write a function that parses CSV files" with "write a function that parses CSV files using our existing FileReader abstraction, following the error handling patterns in our utils module, targeting Python 3.11+." The second version gives the model enough to work inside your system instead of inventing its own.

Counters: Illusionist of Complexity, Congealing Slop. Reduces the chance that AI output conflicts with your existing codebase

The Constraint Prompt

Explicitly state what the code should *not* do. "Do not introduce new dependencies." "Do not use async unless the caller is already async." "Do not add abstraction layers beyond what's needed for this specific use case."

LLMs are completion engines. They will happily add complexity because complexity appears frequently in their training data, and the only thing standing between that instinct and your codebase is explicit constraint.

Counters: Visionary of Grandiose Plans, Scope Creep Kraken. Prevents AI from expanding scope unprompted

The Skeptic's Follow-Up

After receiving generated code, ask the model to identify weaknesses in its own output. "What are the failure modes of this approach?" "What edge cases does this not handle?" "What assumptions does this code make that might not hold in production?"

This works not because the AI's self-critique is always accurate, but because the exercise forces *you* to think critically about the output instead of accepting it at face value.

Counters: Faithful Automator, Probability Pixie. Breaks the cycle of uncritical acceptance

The Provenance Check

When AI generates code that uses a specific library, pattern, or approach, ask: "Where does this pattern come from? Is this a common approach or something specific to a particular codebase?" This won't catch everything, but it creates a habit of questioning origins.

Counters: Bandit of Borrowed Brilliance, Outlaw of AI Plunder. Introduces friction before potentially problematic code enters your codebase

Validating What the Model Gives You

Start with verify. Work toward trust from there.

Staged Acceptance

Never accept AI-generated code directly into your main branch. Use a workflow with explicit gates:

1. **Generate**: AI produces candidate code
2. **Review**: A human reads and understands the code (the "Explain It" rule)
3. **Test**: Code is tested against requirements, including edge cases the AI may not have considered
4. **Integrate**: Code is merged only after passing review and testing

There is nothing radical here. It is what good teams already do for human-written code. The intervention is making it explicit for AI-generated code, because AI output passes the visual sniff test that sloppy human code often fails. It looks finished before anyone has actually checked it, and that appearance is exactly what makes skipping steps feel safe.

Counters: Bulldozer of Confidence, Daredevil of Disaster, Phantom Intern

The Diff Audit

Periodically audit your version control history. What percentage of merged code was AI-generated? What's the defect rate in AI-generated vs. human-written code? What modules have the highest concentration of AI-generated changes?

Without the numbers, the accumulation is invisible. If 80% of your recent commits are AI-generated and your defect rate is climbing, you have found a correlation worth investigating.

Counters: Congealing Slop, Temporal Tangle Worm. Makes the accumulation of AI-generated code visible

The Rewrite Test

For critical code paths, try this: take the AI-generated implementation, delete it, and write the same feature by hand. Compare the two. Where do they differ? Is the AI version genuinely better, or just different? This is expensive, so reserve it for code that handles money, security, data integrity, or user safety.

Counters: Faithful Automator, Illusionist of Complexity. Recalibrates your sense of what the AI is actually contributing

Team Retrospective Questions

Retrospectives are where patterns become visible, if you ask the right questions. Add these to your regular retro rotation. You don't need all of them every sprint. Pick two or three that match your current concerns.

Awareness Questions

- "Did we accept any AI-generated code this sprint that we don't fully understand?"
- "Were there moments where we deferred to the AI's judgment instead of our own? Were those the right calls?"
- "Did our scope change because of AI-generated suggestions? Was that change intentional?"

Surfaces: Faithful Automator, Bulldozer of Confidence, Scope Creep Kraken

Quality Questions

- "What's our ratio of AI-generated to human-written code this sprint? Is that ratio intentional?"
- "Did any bugs this sprint trace back to AI-generated code? What was the root cause?"
- "Are there areas of the codebase where nobody feels confident making changes? Why?"

Surfaces: Congealing Slop, Phantom Intern, Shadow Patcher of Ephemeral Solutions

Process Questions

- "Are we using AI tools because they help, or because we've been told to?"

- "What ceremonies or processes did we follow this sprint that didn't produce value?"
- "Are we archiving or logging AI interactions that nobody will ever review?"

Surfaces: Ritualist of Redundancy, Doomsday Archivist, Stagnant Miasma

Strategic Questions

- "If we turned off all AI tools tomorrow, which parts of our workflow would break? Is that acceptable?"
- "Are we building skills or building dependencies?"
- "What did we learn this sprint that we didn't know before, from our own work, not from AI output?"

Surfaces: Shield Bearer of Stagnation, Stagnant Miasma, Archmage of Analysis Paralysis

Organizational Interventions

Some patterns survive every team-level intervention because the conditions producing them are organizational. Fixing those requires support from above the team.

AI Usage Policies That Actually Work

Most corporate AI policies are either too vague ("use AI responsibly") or too restrictive ("don't use AI at all"). Neither works. The policies that actually change behavior are specific enough that a developer can read one and know, without asking a manager, whether what they are about to do is inside the lines:

- Which AI tools are approved and for which tasks
- How code provenance and attribution get tracked
- When AI-generated content must be disclosed in code reviews
- What level of human understanding is expected for all shipped code
- How often the policy itself gets revisited, because the tools will change faster than the policy cycle

Counters: Outlaw of AI Plunder, Faithful Automator. Gives teams clear boundaries instead of ambiguous expectations

Skill Maintenance Programs

AI tools atrophy the skills they automate. If your team stops writing SQL by hand because the AI does it, eventually nobody can evaluate whether the AI's SQL is good. Budget time for deliberate skill practice, including pair programming without AI tools, manual debugging sessions, architecture exercises.

Counters: Shield Bearer of Stagnation, Stagnant Miasma. Keeps human skills sharp even as AI handles routine work

The AI Budget

Treat AI tool usage like a budget, not a buffet. Not every task benefits from AI assistance. Explicitly identify which tasks are good candidates (boilerplate, repetitive transformations, exploration) and which should remain human-driven (security-sensitive code, architectural decisions, code that handles financial transactions).

Counters: Daredevil of Disaster, Visionary of Grandiose Plans. Prevents AI from being applied indiscriminately to everything

Picking Your Interventions

You don't need all of these. You need the ones that match your symptoms. Use the Diagnostic Field Reference at the start of this chapter to identify which patterns are active in your environment, then pick the corresponding interventions. Start with one or two. Give them a sprint or two to settle in. Add more only when the first ones have become habit.

Aim for sustainable discipline, not a compliance checklist. If an intervention is not producing value, drop it and try something else. The anti-patterns will keep surfacing, the same tendencies, different sprints, different tools. Your counter-moves are the part that should get sharper over time.

Conclusion: Build Better Habits, Not Better Excuses

A small guild-like team studies the work together while trouble waits beyond the doorway.

If you made it this far, you now have something most teams do not: a shared vocabulary for the ways AI-assisted coding goes wrong.

These aren't labels for *people*. They're labels for *behaviors*, and that distinction matters. Your best engineer can exhibit Bulldozer tendencies on a deadline-pressured Tuesday and catch themselves by Wednesday. The whole point of naming the patterns is making the catching easier.

That idea, that naming a pattern gives you power over it, is older than this book and older than AI. The Gang of Four gave us pattern names so we could stop re-explaining the same designs in every meeting. Brown and colleagues gave us anti-pattern names so we could diagnose dysfunction instead of merely complaining about it. This field guide tries to do the same thing for the specific mess we are living through right now.

The RPG frame was never decoration. Classes, artifacts, and monsters gave the patterns shape and humor, but the real purpose was durability. “We have a Scope

Creep Kraken problem" is a sentence that survives a standup. "I think our velocity numbers are misleading because we are generating more code than we are understanding" is a paragraph that gets lost before anyone finishes it.

I should be honest about something. I wrote this book because I recognized the patterns on my own teams and in my own work, not because I was safely observing from the outside. I have been the Daredevil more times than I would prefer to admit. I have watched slop congeal in codebases I was responsible for and told myself the cleanup could wait. The taxonomy in this book is not a set of accusations I am leveling at other people. It is a field guide I need as much as anyone.

If there is one idea worth keeping from the whole framework, it is this: our familiar strengths age into familiar excuses much faster when AI is in the loop. The person who gets things shipped can start mistaking motion for judgment. The person who sees the system clearly can keep planning until the quarter is gone. The person who finds every shortcut can leave behind work nobody else can safely inherit. The person who builds the process can start trusting the process more than the software.

Here is what I hope you do with it.

Use the names. In code reviews, in retros, in planning sessions. When you say "I think we're feeding the Scope Creep Kraken here" and your whole team knows what that means, you've compressed a five-minute debate into one sentence.

Start with yourself. The most uncomfortable chapter in this book should be the one where you recognized your own habits. Run through the diagnostic field reference in the previous chapter and be honest. The patterns you are most reluctant to admit are usually the ones worth naming first.

Bring it to your team. Pick two or three patterns that fit your current situation, discuss them honestly, and adapt them to fit the way your team actually works. If a name does not fit, rename it. The point is the conversation, not the catalog.

Stay skeptical of the tools and of yourself. AI coding tools are genuinely useful, and they are also genuinely capable of making you worse at your job if you let them. The question was never simply whether to use AI, but whether you are using it in a way that keeps you sharp or in a way that lets you quietly stop being one.

Review like it matters. Treat AI-generated code the way you would treat a junior developer's first PR, with rigor, with patience, and with the expectation that you will need to teach it something. When something feels off but you cannot quite name it yet, trust that feeling. It is usually right. And be honest about the scale: you may not be able to review every line of code your agents produce, but you must stay connected to the design decisions, the architecture, and the patterns in what you ship. "Vibe coding" is a fine philosophy for weekend projects. For production software, these tools still need a responsible engineer behind them. That is not going to change.

The tools will keep changing, and the models will get better. Some of the specific failure modes in this book will fade as the technology matures. I hope the Probability Pixie gets less dangerous, and I suspect the Phantom Intern will be with us for a long time. But the underlying dynamics are older than this moment: over-trust in automation, the appeal of speed over understanding, and the slow erosion of skills you stop practicing. Those were true before AI, and they will be true after whatever comes next.

You've got the names and the tools. Now comes the hard part: actually using them, every day.

Acknowledgements

My wife Susan, my son Thomas, and my daughter Josephine deserve my first thanks for being patient with me while I disappeared for a few weekends to write this book.

I also want to recognize Chris Undernehr as a Contributor for his significant contributions to the second edition of Volume I, and to thank Steve Sullivan for helping me work through some of the ideas in the initial brainstorming.

I am also grateful to Colin Henry and Oakley Hall for providing feedback.

I also want to thank Mike Loukides for helping me avoid publishing a book with an unsellable title. The initial title was *Slop Codex*, and I changed it because of his feedback.

A Note on Methodology

This book is not academic research. It does not have a sample size, a control group, or a peer-reviewed methodology section. What it has is observation — years of it, across enough teams, companies, and technology stacks that the patterns started repeating with uncomfortable regularity.

The failure modes described here emerged from my own experience building and managing software teams, from conversations with other engineering leaders navigating the same transition, and from watching the same dynamics play out independently in organizations that share nothing except the underlying pressures of AI-assisted development. Some of the patterns crystallized during specific projects. Others accumulated gradually as I noticed the same sequence of decisions leading to the same category of trouble in contexts that had no connection to each other.

I did not set out to build a taxonomy. I set out to name the things I kept seeing, because naming them made them easier to talk about. The RPG framework came later, as a way to make the names stick and the relationships between patterns visible. The framework is a compression tool, not a measurement instrument.

Where the book makes claims about prevalence — that certain patterns are common, that they appear across different team sizes and tech stacks — those claims are based on pattern recognition, not quantitative measurement. I have tried to be honest about that throughout. When I say "I keep seeing this," I mean it literally: I keep seeing it. That is the evidence. It is the kind of evidence that practitioners have always used to build shared vocabulary, from Christopher Alexander's pattern language to the Gang of Four's design patterns to Brown's anti-patterns catalog. None of those books came with datasets either. They came with recognition — the experience of reading a description and thinking, "I know exactly what that is."

If the patterns in this book match your experience, they are doing their job. If they do not, set them aside and name the patterns you are actually seeing. The vocabulary matters more than my particular version of it.

Glossary

Canonical definitions for every named entity in the *AI Developer's Field Guide.* Use this as a quick reference when someone mentions a term in a meeting and you need the two-sentence version.

Classes

Fighter. The force-first archetype. Fighters use AI tools to push through problems with speed and certainty, generating code at volume and merging with confidence — often without the review, testing, or understanding that would justify that confidence. Their failure mode is collateral damage: things break because speed outpaced caution.

Wizard. The abstraction-first archetype. Wizards use AI tools to analyze, compare, and architect — endlessly. They build elaborate frameworks, request comprehensive comparisons, and design for hypothetical futures. Their failure mode is paralysis and unnecessary complexity: the system grows more sophisticated while actual delivery stalls.

Rogue. The speed-and-opportunity archetype. Rogues use AI tools to move fast, grab solutions from wherever they can, and patch problems just enough to move on to the next thing. They ship quickly, but the code is brittle, the provenance is unclear, and the shortcuts create liabilities. Their failure mode is accumulated fragility and integrity risk.

Cleric. The ritual-and-faith archetype. Clerics follow process with devotion and trust AI output with the certainty of doctrine. They build elaborate ceremonial workflows around AI tools, log everything, automate everything, and question nothing. Their failure mode is process gravity: the rituals continue long after they've stopped producing value.

Anti-Patterns

Fighter Anti-Patterns

Bulldozer of Confidence. The developer who generates AI code at volume and merges it with minimal review, treating AI output as pre-validated. Characterized by high velocity metrics, low comprehension of shipped code, and pull requests that move from generation to production with insufficient friction. The Debugging Greatsword is their weapon — they brute-force through problems by regenerating code until something works, rather than understanding why something failed.

Shield Bearer of Stagnation. The developer or team that rejects AI tools and new approaches on principle, clinging to established patterns not because they're better but because they're familiar. They wield the Design Patterns Relic — invoking classic software patterns as a defense against any tool or practice that didn't exist when they learned to code. Their conservatism isn't wisdom; it's fear wearing wisdom's armor.

Daredevil of Disaster. The developer who applies AI tools to high-risk, high-consequence areas — infrastructure, security, financial logic — without the domain-specific review those areas demand. They carry the Unapproved Gauntlet of Power, wielding AI capabilities in contexts where the cost of a subtle error is catastrophic. Fast and reckless where they should be slow and careful.

Wizard Anti-Patterns

Archmage of Analysis Paralysis. The developer or architect who uses AI tools to endlessly research, compare, and evaluate options without ever committing to a decision. They consult the Scroll of Borrowed Fluency — generating comparison after comparison, analysis after analysis, seeking perfect certainty in a domain where it doesn't exist. Sprints pass with no code written while the analysis documents grow ever longer.

Illusionist of Complexity. The developer who uses AI tools to generate architectures far more complex than the problem requires. They carry the Bag of arXiv Ontologies — building elaborate category systems, abstraction hierarchies, and design patterns that serve no current requirement. Their code compiles, passes review (because reviewers can't tell if the complexity is justified), and creates a maintenance burden that compounds over time.

Visionary of Grandiose Plans. The developer or architect who uses AI tools to plan for a future that hasn't arrived — and may never arrive. They brandish the Scepter of Infinite Acceleration, designing systems for millions of users when they have hundreds, building platform capabilities before the first product ships, and optimizing for scale that exists only in their projections. The AI enables their vision by producing plausible-looking designs for any hypothetical scenario they can describe.

Rogue Anti-Patterns

Shadow Patcher of Ephemeral Solutions. The developer who uses AI tools to patch symptoms rather than fix causes. They wear the Cloak of RAG, generating surface-level fixes that make the immediate problem disappear while leaving the underlying defect intact. Their patches pass tests (often because the tests are equally superficial) and the same bugs resurface in different forms the following sprint.

Bandit of Borrowed Brilliance. The developer who accepts AI-generated code without understanding its origins, approach, or implications. They wear the Mask of Stolen Insights — submitting code they didn't write and can't explain, using patterns and libraries they haven't evaluated, building on solutions whose provenance is unknown. When the code fails, they can't debug it because they never understood it.

Outlaw of AI Plunder. The developer who ignores the intellectual property, licensing, and attribution implications of AI-generated code. They carry the Dagger of Terms Violation — introducing copyleft-contaminated code, reproducing proprietary patterns, and violating tool terms of service without awareness or concern. The risk is legal and reputational, and it compounds silently until an audit surfaces it.

Cleric Anti-Patterns

Faithful Automator. The developer who trusts AI output the way a believer trusts scripture — without question, without verification, without the skepticism that good engineering demands. They wear the Medallion of Unquestioning Trust, treating AI-generated code, tests, and recommendations as authoritative. When AI output is wrong, they don't catch it because they've stopped looking.

Ritualist of Redundancy. The developer or team that builds elaborate ceremonial processes around AI tools — daily AI-generated status reports, automated AI summaries of meetings, AI-assisted code reviews of AI-generated code — without asking whether any of it produces value. They venerate the Idol of Eternal Standups, performing process for the sake of process, generating output that nobody reads and artifacts that nobody uses.

Doomsday Archivist. The developer or team that hoards AI-generated output, logs, conversations, and artifacts with the conviction that they might be needed someday. They carry the Eternal Backup Bolt, archiving everything and deleting nothing. Storage costs climb, data governance becomes impossible, and the archive grows into a liability disguised as an asset.

Artifacts

Debugging Greatsword. The Fighter artifact associated with the Bulldozer of Confidence. Represents the practice of brute-force debugging through repeated AI code generation rather than systematic diagnosis. Powerful in the hands of someone who understands the problem; destructive in the hands of someone who doesn't.

Unapproved Gauntlet of Power. The Fighter artifact associated with the Daredevil of Disaster. Represents the use of AI tools in high-stakes contexts without appropriate authorization, review, or domain expertise. The gauntlet grants power; it does not grant judgment.

Design Patterns Relic. The Fighter artifact associated with the Shield Bearer of Stagnation. Represents the invocation of classic software engineering patterns as a defense against new tools and approaches. The patterns themselves aren't the problem — using them as a shield against all change is.

Scroll of Borrowed Fluency. The Wizard artifact associated with the Archmage of Analysis Paralysis. Represents the endless generation of AI-powered research, comparisons, and analyses that defer decisions indefinitely. The scroll always has one more section to consult.

Bag of arXiv Ontologies. The Wizard artifact associated with the Illusionist of Complexity. Represents the accumulation of unnecessary abstraction layers, taxonomy systems, and architectural complexity generated by AI tools. The bag always has room for one more abstraction.

Scepter of Infinite Acceleration. The Wizard artifact associated with the Visionary of Grandiose Plans. Represents the AI-enabled practice of designing for hypothetical scale and future requirements that don't yet exist. The scepter promises infinite growth; it delivers premature optimization.

Cloak of RAG. The Rogue artifact associated with the Shadow Patcher of Ephemeral Solutions. Represents the use of retrieval-augmented generation and similar AI techniques to produce patches that address surface symptoms while concealing deeper defects. The cloak makes problems invisible; it does not make them absent.

Mask of Stolen Insights. The Rogue artifact associated with the Bandit of Borrowed Brilliance. Represents AI-generated code accepted and submitted without understanding its origins, approach, or reasoning. The mask lets you present borrowed knowledge as your own — until someone asks you to explain it.

Dagger of Terms Violation. The Rogue artifact associated with the Outlaw of AI Plunder. Represents the legal and ethical risks introduced when AI-generated code carries licensing obligations, proprietary patterns, or terms-of-service violations. The dagger cuts both ways: it solves the immediate problem and creates a future liability.

Medallion of Unquestioning Trust. The Cleric artifact associated with the Faithful Automator. Represents the practice of accepting AI output as authoritative without human verification or critical review. The medallion inspires faith; it does not inspire accuracy.

Idol of Eternal Stand-ups. The Cleric artifact associated with the Ritualist of Redundancy. Represents ceremonial processes built around AI tools that persist regardless of whether they produce value. The idol demands worship; it does not demand results.

Eternal Backup Bolt. The Cleric artifact associated with the Doomsday Archivist. Represents the compulsive archiving of AI-generated outputs, logs, and artifacts without a strategy for when or whether they'll ever be used. The bolt secures everything and releases nothing.

Monsters

Scope Creep Kraken. The environmental monster that emerges when AI tools make it trivially easy to generate new features, expand requirements, and imagine capabilities that nobody asked for. The Kraken grows one tentacle at a time — each addition seems reasonable in isolation, but the cumulative effect is a project that collapses under its own expanding scope. Fed by Visionaries, tolerated by Bulldozers, invisible until it's too late to cut back.

Congealing Slop. The environmental monster that forms when AI-generated code accumulates without consistent standards, patterns, or human understanding. The Slop isn't any single piece of bad code — it's the aggregate. Inconsistent styles, conflicting patterns, redundant implementations, and dead code gradually merge into a codebase that resists comprehension and refactoring. The longer it congeals, the harder it is to separate the functional from the vestigial.

Probability Pixie. The environmental monster that appears when teams treat probabilistic AI output as if it were deterministic. AI-generated code looks correct — variable names are sensible, control flow appears sound — but carries a baseline error rate that compounds when output feeds into further generation. The Pixie's gift is a world where 94% accuracy feels like 100% until you're standing in the 6% that's wrong. The danger isn't dramatic failure; it's subtle wrongness that erodes trust in the codebase.

Phantom Intern. The environmental monster that manifests when AI-generated code persists in a codebase after the context for understanding it is gone. The Phantom Intern is the invisible contributor who left no documentation, attended no meetings, and can't answer questions. Their code works — until it doesn't — and when it breaks, nobody on the team can diagnose it because nobody on the team wrote it or understands why it was written that way.

Stagnant Miasma. The environmental monster that settles over teams that use AI tools exclusively to replicate existing approaches, never to explore new ones. The Miasma is the death of innovation through optimization of the familiar. Every solution looks the same because the AI was trained on the same patterns, and the team has stopped questioning whether those patterns are the right ones. Creativity atrophies. The team ships functional, conventional, and increasingly irrelevant software.

Temporal Tangle Worm. The environmental monster that lives in the roadmap, coiled around past decisions the team cannot undo and future promises it has not earned. The Worm appears when sensible foresight turns into devotion to a future version of the product or model stack that does not exist yet. Its core trick is making tomorrow's possibilities feel like today's obligations — deferring present work because the next model release will supposedly make it obsolete. The Worm feeds on sunk costs, demo-driven development, and the gap between where the team is and where it imagines it will be.

Index

www.ingramcontent.com/pod-product-compliance
Lightning Source LLC
LaVergne TN
LVHW081324110826
845149LV00007B/1580

* 9 7 9 8 9 9 5 2 5 9 7 4 9 *